STUDIES FOR SMALL GROUPS

The JEHOVAH'S WITNESSES

UNDERSTANDING THEIR FAITH AND TEACHINGS

Michael C. McKenzie, Ph.D.

Copyright © 1999
College Press Publishing Co.

All Scripture quotations, unless otherwise noted, are from the NEW AMERICAN STANDARD BIBLE®, © The Lockman Foundation 1960, 1962, 1963, 1968, 1971, 1972, 1973, 1975, 1977. Used by permission.

International Standard Book Number 0-89900-833-X

CONTENTS

Series Introduction 5
Study Introduction 7
1. The Importance of Solid Foundations 11
2. Who Then Is This? 27
3. Salvation, Resurrection, and the Second Coming 46
4. What Jehovah's Witnesses Believe 66
5. Jehovah's Witnesses and Society 86
Appendix: The Early Church Fathers
    and the Deity of Christ 99
Additional Resources 102
Glossary of Terms 105

STUDIES FOR SMALL GROUPS

Welcome to the *Studies for Small Groups* series from College Press. This series is designed for simplicity of use while giving insight into important issues of the Christian life. Some, like the present volume, are topical studies. Others examine a passage of Scripture for the day-to-day lessons we can learn from it.

A number of possible uses could be made of this study. Because there are a limited number of lessons, the format is ideal for new or potential Christians who can begin the study without feeling that they are tied into an overly long commitment. It could also be used for one or two months of weekly studies by a home Bible study group. The series is suitable for individual as well as group study.

Of course, any study is only as good as the effort you put into it. The group leader should study each lesson carefully before the group study session, and if possible, come up with additional Scriptures and other supporting material. Although study questions are provided for each lesson, it would also be helpful if the leader can add his or her own questions.

Neither is it necessary to complete a full lesson in one class period. If the discussion is going well, don't feel that you have to cut it off to fit time constraints, as long as the discussion is related to the topic and not off on side issues.

College Press is happy to add this new 5-lesson study on the Jehovah's Witnesses to the *Studies for Small Groups* series. It is our hope that this will be helpful in understanding these zealous workers with a misguided message. Look for other studies on non-Christian and pseudo-Christian faiths.

THE JEHOVAH'S WITNESSES

You've had one of those days at work. After fighting rush hour traffic, you finally pull into the driveway, utterly exhausted. As you collapse into your favorite recliner, kick off your shoes and finally begin to relax, the doorbell rings. Dead tired and annoyed, you shuffle to the door.

"We're Bible students," a perky mother with two young girls announces. "May we come in to talk to you about the word of Jehovah?"

You don't know exactly who they are, but you've seen and heard enough for one day. "Look," you sputter through clenched teeth. "My church warned us about you guys, why can't you just go away and mind your own business?" And before any of your visitors can reply, you shut the door . . . hard. As you walk grimly back to your chair, you tell yourself, *Serves 'em right. Who do they think they are, bothering people like that?* Still, you can't shake the image of one smiling mother, two young girls, and two older men — as the door slammed, they all almost seemed to *expect* it. *Maybe I was too hard on them. After all, all they wanted to do was talk . . .*

Most of us have had some experience like this. More often than not, we are able to identify the visitors as Jehovah's Witnesses, even though they may start the conversation by calling themselves "Bible Students," or "Interested Students of Religion." Many of us have also been frustrated after talking with Jehovah's Witnesses. Not only are they earnest, but they seem capable of quoting the Bible endlessly! And, speaking of Bibles, what about their claim that the Christian church tainted our modern translations, forcing them to produce a new and more accurate translation? However, many people are simply frustrated by their tactics. Also, Jehovah's Witnesses seem to agree with much of what the Christian says, but afterwards, the Christian is deeply uneasy about such "agreement." What's going on?

Because Jehovah's Witnesses are so active in their communities, their influence often extends far beyond their numbers. They actively pursue converts door to door, often causing Christians to wonder, "Should I be doing that?" Because of similarities in their doctrines to orthodox Christianity, and their vigorous campaigns of neighbor canvassing, there is a need for an easy-to-understand text that compares the faith of Jehovah's Witnesses to historical, orthodox Christianity. Thus, the idea for this book was born. It is my hope that this work will be used by Christians so that they not only have a better grasp of *what* they believe, but come to understand the *foundation* for their own beliefs. In addition, as they come to understand the beliefs of Jehovah's Witnesses, Christian readers will better understand how to *defend* what they believe.

One of my mentors in graduate school, Walter Martin, was fond of telling the story of how many banks and financial institutions trained their employees to spot counterfeit money. For weeks on end, the tellers would handle nothing but the real thing. They would touch it, feel it, smell it — become intimately familiar with it. Then, and only then would they be exposed to the phony money. Inevitably, the tellers would then easily detect the counterfeit currency — not because they were experts regarding the marks of a fake bill, but *because they knew exactly what the real thing was like*. So it is with Christianity. It is essential for the reader to become intimately familiar with the doctrines, the people, the history of historical, orthodox Christianity. Then, he or she will easily be able to detect the counterfeit.

I have deliberately designed this work to be "user-friendly." That is, this book is written at the popular level, with a bare minimum of academic language. I don't mean that the book is simplistic, or that it skimps on scholarship. Rather, it is designed to be read by people from all walks of life, from formal classrooms to informal living rooms. The various questions at the end of each chapter are designed for the reader to test his or her own knowledge before moving on. I have also placed additional readings at the end of each chapter, so the reader can go beyond the text for further learning. In addition, I have included a glossary of terms at the end of the work. Theology has developed its own language, and at times, it's important to "translate" such language into common English. Lastly, I have included some helpful groups and organizations in the closing "Additional Sources." Readers thus have ample *practical* resources to aid their own education. Churches and other groups will find this book helpful both in their Christian education programs and for their libraries. Individuals can also use it to work through at their own pace. No matter the context, it will hopefully ignite in you a desire to learn more about your faith, and encourage you in your own spiritual journey. It is certainly *not* designed to poke fun or embarrass Jehovah's Witnesses — that sort of attitude has no place in our discussion.

Because of the nature of the work, however, this book does focus on the points of disagreement between Jehovah's Witnesses and historical Christianity. Especially, it focuses on the person and mission of Jesus. Who was he? Why did he come to earth? Given the importance that historical Christianity has traditionally placed on such questions, and given the fact that Jesus is the central person in Christianity, it is absolutely essential to examine competing answers to these tough questions. It is one thing to admit that different Christian denominations differ on matters more peripheral to Christian faith, it is entirely another when one group has a radically different view of the founder of the faith. When such is the case, it is no longer a matter of "minor family squabbles," but an understanding that the group in question has placed itself outside the boundaries of historical Christianity.

I had been a Christian only for a short time before I knew I wanted to learn more about the Christian faith. I was fortunate to begin my study at Simon Greenleaf School of Law, studying under such teachers as Walter Martin. Martin, a recognized expert in religious sects and groups in America, made a lasting impression on me. As I mentioned, he never wavered in his insistence that the best way to defend Christianity against attack was to *know* Christianity. I have always found that maxim to be true.

I am thankful to John Hunter at College Press for his patience and encouragement. I am also thankful to my parents for raising me in a home that both nourished my faith, and encouraged me to find answers to tough problems. Lastly, I am thankful to my wife Allison who has supported me through the tough times.

ONE

THE IMPORTANCE OF SOLID FOUNDATIONS

When I was a boy, one of my favorite pastimes was to explore new construction sites. My friends and I would watch for hours as a house or other building magically took shape before our very eyes. It seemed like a giant erector set had come to life! I'll never forget the time we rode our bikes to a new site not far from our neighborhood. Construction was just beginning, and the primary evidence of their work was a large hole in the ground, with mounds of dark brown earth thrown up around the edges. What a discovery! It was always a find to locate a site so new. We'd make friends with the workers, sometimes sharing their lunches, and every now and then, as our reward, we'd be treated to displays of colorful language seldom heard at home. We were convinced our help was indispensable to the carpenters, bricklayers, and electricians. We were truly in heaven.

It didn't take long before things started to take shape in the hole. Masons and bricklayers arrived — soon followed by the vehicle that occupies a place dear to the heart of any seven-year-old: a cement mixer truck. We watched in awe as they poured the foundation, roarings, grindings, and the lime-like smell of cement filling the air. Being "experts" in construction, we knew that soon the walls and floors would make their appearance.

When we next arrived at the site, we skidded our bikes to a stop, nearly falling into what was once again, only a very large — and empty — hole. A solitary man was cleaning up the bottom of the pit, scraping up hunks of cement into a wheelbarrow.

"Hey," we shouted, almost in unison, "What happened to the rest of it?"

The burly workman stopped what he was doing and peered at us, recognition slowly dawning. "We had to tear it down. It wasn't up to specs." Noting the blank looks on our faces, he smiled as he translated, "The foundation was cracked and crooked, a big wind or storm might have blown the house down."

I still remember how sad we were that something that seemed so perfect had to be torn down. As an adult, however, I can sympathize with the homeowners and their concern to get things right. Without a proper foundation, the rest of the house would literally be on borrowed time. So it is with philosophy or theology. When one is constructing a worldview, a philosophy of life that makes sense of the world, it is essential to "get things right." *One must indeed build on solid foundations if the "building" is to last.* I suppose that's why the words of Jesus regarding the man who built his house upon the rock made sense to me early on — because his foundation was solid, his house could withstand the storms and trials of life (Luke 6:46-49). Thus, we need to know right away what's absolutely essential about Christianity. What is our "foundation"; what must we believe in order to be called "Christian?" To illustrate what's foundational about Christianity, let's take a brief look at three biblical passages.

These passages are foundational because each goes to the very *heart* of Christianity. If one looks in the yellow pages under "Churches," he or she is immediately struck by the seemingly endless diversity of denominations. Many churches disagree on the mode of baptism, whether one should (or can) speak in "tongues," the importance of

> **One must indeed build on solid foundations if the "building" is to last.**

images inside the church building, how the minister should dress, the issue of women's ordination, the question of whether believers should drink alcohol, and so on. But, if a person digs more deeply, he or she would find a good deal of agreement on what might be called the "cardinal doctrines" of Christianity. To define what such doctrines might be, think of those beliefs that one *must* believe in order to be called Christian.

> The New Testament is specific in warning us about "false Christs and prophets"

Even the newest Christian probably knows that such beliefs have something to do with believing in Jesus. But what does it mean to "believe *in*" Jesus? Does the *identity* of Jesus matter at all? Isn't it more important to simply believe — at least in something? Well, *the New Testament is specific in warning us about "false Christs and prophets"* (Matt. 24:24), a false gospel (Gal. 1:6), false apostles (2 Cor. 11:13), and false brethren (Gal. 2:4). In other words, the New Testament says it matters a great deal that the *object* of one's faith and trust is true and deserving of that trust. Mere belief is not enough; faith can be wrongly directed.

This brings us to the difference between *objective* and *subjective* truth. Subjective truth has to do with truth claims that are true, depending on the *subject*. For example, if you commented to a friend, "My, that shirt looks great on you," such a truth claim is totally dependent on the subject wearing the shirt. The same shirt on somebody else might look horrible. *Objective* truth, however, has to do with the object of the claim, and is true whether you believe it or not. For example, I might refuse to believe that Olympia is the capital of Washington — perhaps I believe it's Seattle. So much the worse for me! No matter how hard I hold to my belief, I'm still wrong. Let me illustrate how objective truth works. Suppose someone believes that an airplane with only one good wing will carry him safely over the mountains. One may believe such a thing with all of his being — but since the *object* of his belief is false, all the belief in the world won't save him. Since the Bible warns so clearly of *false* objects of belief, believing wrongly must be possible, and believing correctly must be essential. *Scripture thus clearly holds to an **objective** view of truth when it comes to the essentials of the faith.*

I chose three biblical passages as foundational — not because they are the only ones that might fit the bill — but because they are among the *clearest* that do. One concerns the question of a man who wants to know — right up front — what is required to "make it to heaven." We might be offended at his boldness — but Paul gives him a succinct answer to a succinct question. Our next passage concerns the claims of Jesus himself. Some say he claimed to be no more than a simple teacher. In our passage, however, the words of Jesus nearly get him killed. What do his words tell us about the identity and mission of Jesus? The last passage concerns the amazing faith of the thief on the cross. He sees in Jesus someone special, special enough to put his own last-chance faith in. Although we humans might be skeptical of such last-second conversions, *Jesus responds to the thief, not in theological jargon or in cynicism, but in clear words impossible to misunderstand*, a direct promise of heaven. All three passages then, reveal the very heart of Christianity, the foundational doctrines of how to get to heaven, and how Jesus might fit into it all.

WHAT'S THE BOTTOM LINE?

As you read and examine these passages, think of others in Scripture that you consider foundational — what characteristic(s) do they share?

Acts 16:30-31. "Sirs, what must I do to be saved?" And they said, "Believe in the Lord Jesus, and you shall be saved, you and your household."

> Scripture clearly holds to an *objective* view of truth when it comes to the essentials of the faith.

A classic foundational passage. The Philippian jailer is aroused by the earthquake that destroyed the jail which held Paul and Silas. Supposing his prisoners had all escaped, the official prepares to kill himself, knowing that since he was responsible for the prisoners' security, his own life is now forfeit. Paul and Silas reassure him, and the jailer somehow intuits that they are responsible for the supernatural release of the prisoners. The jailer now knows that these men are indeed God's spokesmen (read Acts 16:16-18), and trembles in fear

> Jesus responds to the thief in clear words impossible to misunderstand.

before them. He wants to know what exactly is required for salvation.

At first glance it might seem that all the Apostle Paul requires is simple belief in the existence of Jesus, a simple acknowledgment that he walked among men. Such a reading, however, will not stand up under closer scrutiny. Later in the Scriptures, when the Apostle James discusses faith and belief, he admits that those "do well" who believe in God, but goes on to say that even the demons "believe" in the "fact" of God's existence (Jas. 2:19). Since it is clear that demons are not Christian, and have no hope of becoming so, we are forced to take another look at Paul's words to the Philippian jailer in Acts. What does he mean by "believe"?

First, when Paul replies to the jailer that he must "believe," the Greek word used (*pisteuo*) suggests far more than mere agreement about certain facts (for example, that Sacramento is the capital of California). Rather, its close linguistic relationship to the Greek word for faith (*pistis*) or for trustworthy (*pistos*) suggests that the belief that Paul speaks about means something like "have trust in," or "trust."[1] This fits exactly with James's warning that Christians' belief about God must be more than simple agreement about facts. The jailer's immediate willingness to be baptized at that same hour of the night shows that his faith was the kind that necessitates action.

This way of looking at belief also fits with the *object* of belief in this passage. When answering the jailer, Paul links this belief with "the Lord Jesus," and this title for Jesus points to the importance of *identity* of the object of belief. It is not simple assent in Jesus' existence, nor is it even trust in a mere man Jesus; Paul's answer says that it is trust in the *Lord* Jesus that's all important. *For a Jewish Rabbi like Paul to call Jesus "Lord" in this context is to give him the same honor, glory, and power that is due only to God.* Indeed, to call Jesus "Lord" is to confess him as ruler over all, as one who rules with "all things in subjection under his feet" (Eph. 1:22).[2]

Paul's words to the jailer now begin to take on new meaning. We might suspect that this Jesus, whom Paul calls "Lord," is no ordinary Jewish rabbi. Thus, when Paul commands belief in the Lord Jesus as a requirement for salvation, we are not surprised to find that this belief in Jesus is somehow equated shortly after by Luke (the author of Acts) as belief in *God* (Acts 16:34). To trust in Jesus is to trust in God. Thus, we need to take a closer look at this man Jesus.

John 8:24,58. "'I [Jesus] said therefore to you, that you shall die in your sins; for unless you believe that I am *he*,[3] you shall die in your sins.' . . . Jesus said to them, 'Truly, truly I say to you, before Abraham was born, I am.'"

When one reads the Gospels, there are times it seems as if Jesus is in one debate after another with the Pharisees, the Jewish religious leaders of the day. Nowhere are these debates sharper than in the Gospel of John. In our passage in question, Jesus points to his own identity as proof of the truth of his mission; since the Pharisees oppose Jesus, they are, in effect, opposing his mission and, by inference, God. In chapter 8, matters have finally come to a head.

Immediately after the account of the woman caught in adultery (which ends in verse 11), Jesus claims boldly that he is "the light of the world," and says that those who follow him will have "the light of life" (v. 12). A daring claim indeed. The Pharisees immediately accuse him of pride, pointing out that Jesus was, in effect, "blowing his own horn" about his mission and ministry (v. 13). Jesus then replies by linking his claims to his relationship with his Father: the Father had testified in his behalf, Jesus' voice was not alone (v. 18).

> **For a Jewish Rabbi like Paul to call Jesus "Lord" is to give him the same honor, glory, and power that is due only to God.**

At this point Jesus obviously is not overly concerned with making friends. He tells the Pharisees that they will "die in their sins" because they were "from below" while he was "from above" (vv. 21, 23). In essence, Jesus is saying, they will die with unforgiven sins. In modern terms, not only is the debate getting "personal," but Jesus is clearly spotlighting a huge divide between himself and the Pharisees — God's judgment is now upon them. For their part, the Pharisees are obviously beside themselves with rage, going so far as to think Jesus might even resort to suicide, since he was going to a place where they "could not follow" (v. 22).

> In the clearest way possible, Jesus links belief in himself, in his *identity*, to God's forgiveness of sins.

The climax of the argument arrives. Jesus repeats that the Pharisees would die in their sins, but now adds the *reason* that they were in this sorry state: because their view of Jesus was false: "For unless you believe that I am *he*, you shall die in your sins" (v. 24). Any orthodox Jew knew what Jesus meant by "dying in one's sins." It meant to lose all hope of redemption, to give up any thought of dwelling with God in eternal fellowship with the patriarchs of old. Thus, *in the clearest way possible, Jesus links belief in himself, in his **identity**, to God's forgiveness of sins*. But what exactly is Jesus getting at here? What sort of identity is he claiming?

Most modern translations of the Bible place the "he" in italics, denoting that it is not found in the original Greek text (as for example in *The New American Standard Bible*). Thus, in the original Greek text, the language reads "Unless you believe that *I am*, you shall die in your sins." But who or what is this "I am?" Since Jesus links our having a proper view of his identity to our sins being forgiven, it is absolutely essential to get this right.

For one thing, the Greek original in verse 25 shows clearly that the Pharisees are incredulous when they hear Jesus' claim. The verb tense indicates they then *repeatedly* asked him "Who are you?"[4] In addition, the unnecessary use of the pronoun, and its place far forward in their reply, places emphasis on "you," making their question similar to our modern, "Who do you think *you* are, anyway?"[5] Clearly, Jesus' reply had angered the Pharisees. But why? Let's take a look at the context of verse 58 for help.

To get a proper understanding of what's going on in this verse, we must understand the hallowed place that Abraham occupied in the mind of the pious Jew. Judaism did (and still does) call Abraham "the Father of our religion," and he is seen as the father and example of faith for all believing Jews (as for Christians).

Thus, it's quite understandable that the Jews claim (in verse 33) that they are "Abraham's offspring" and that Abraham is "our father" in verse 39. They are claiming much more than biological heritage; they are claiming that such a heritage gains immediate favor with God. John the Baptist had run into much the same attitude when he began his own ministry. While baptizing believers in the Jordan River, John warned the Jews who came, "Do not begin to say to yourselves, 'We have Abraham for our father,' for I say to you that God is able from these stones to raise up children to Abraham" (Luke 3:8). Thus, *real trust and faith in God was necessary, not merely finding Abraham on one's family tree.*

Jesus understood well this attitude of the Jews toward Abraham. As the conversation between him and the Jews develops (John 8:34-51), Jesus wants the Jews to demonstrate that they are "true" children of Abraham by acting like Abraham. After all, Jesus says, Abraham believed what he heard from God, but the Jews were rejecting it (vv. 39-40, 47). Thus, they had forfeited the right to be called "sons of Abraham." They were mere pretenders.

But Jesus goes even further. He has the audacity to claim that if anybody keeps his word, he would never see death (v. 51)! The Jews undoubtedly now sense victory. After all, beloved Abraham, the father of Judaism had died. So had the prophets. Since these men of God had

> **Real trust and faith in God was necessary, not merely finding Abraham on one's family tree.**

died, *this radical rabbi seemed to claim his identity and message were even greater than those of the Jewish patriarchs* (v. 52). The Jews now sprung the trap: "Surely you are not greater than our father Abraham, who died? The prophets died too; whom do you make yourself out to be?" (v. 53). We can almost see the smug satisfaction on their faces. After all, what can Jesus do? He must either retract his statement about defeating death, or claim something no orthodox Jew ever would: that he was greater than Father Abraham. Jesus, however, isn't finished. Instead of softening his words, his claims and tone become more pointed and more bold.

> This radical rabbi seemed to claim his identity and message were even greater than those of the Jewish patriarchs.

"Your father Abraham rejoiced to see my day, and he saw it and was glad" (v. 56). I'm sure the Jews thought that they were dealing with a lunatic. *Abraham lived thousands of years ago, how could our Blessed Father know anything about this demented Jewish rabbi?* They closed in for the kill. "You are not yet fifty years old, and you have seen Abraham?" (v. 57). The defeat must have seemed total. Incredibly, instead of retreating or backing down, Jesus takes the offensive!

"Truly, truly, I say to you, before Abraham was born, I am" (v. 58). This is by all accounts a truly remarkable claim. Jesus had already claimed to have existed at the time of Abraham (v. 56). That audacious claim had only bewildered and provoked the Jews further. Here, their reaction goes beyond words — the text says "they picked up stones to throw at him." Why did these specific words of Jesus get such a reaction? Why did the Jews want to kill him?

It is doubtful that the Jews were simply so angry that they wanted to kill Jesus in a spontaneous moment of rage. *If they thought that Jesus had merely gone off the deep end, the Pharisees could have laughed off his claims,* using them to weaken his popular support. The fact of their choosing stones to kill him suggests that the Jewish leaders were beginning the Old Testament procedure to execute Jesus. Also, since Jesus was "teaching in the temple" (vv. 2, 20, 59), the Jews would have had plenty of time to cool off as they gathered stones — their deliberate efforts argue again for an attempt at judicial execution. Finally, when John links an attempt to execute Jesus a short time later (10:31), he says that "again" the Jews took up stones, and here the Jews state clearly that they want to execute Jesus for "blasphemy" (v. 33).

The idea that the Pharisees were trying to execute Jesus for blasphemy agrees precisely with how the Jews today view both Talmudic and biblical law. Depending on which source is consulted, Jewish law allows for either 18 or 7 offenses that are punishable by stoning.[6]

The only crime listed that matches the Jewish accusations against Jesus is "blasphemy." Blasphemy itself is defined variously, but the underlying theme is a spurning, a contempt for God that tries to bring Him down to man's level, or, conversely, attempts to raise men up to God's. In either case, humans are refusing to give God the honor due Him — pride and contempt are close bedfellows.[7] As the Jewish understanding of blasphemy evolved, even *to pronounce* God's sacred name — the name of *Yahweh* — was considered blasphemous and worthy of death.[8] Some of this reverence for God's name rubbed off on Christians, and the word commonly translated "Jehovah" comes from substituting the vowels for *adonai* (a, o, a) into the sacred Hebrew name for God, *yhwh*.[9] Let's look again at the passages in question (8:24,58).

In both cases, the original Greek text goes to great lengths to make the point that Jesus was claiming to be "I am" (*egō eimi* in the Greek text). Considering both the Jewish audience and their reaction — incredulity in verse 25 and wanting to execute Jesus in verse 59 — the true import of Jesus' words becomes clear. The Jewish scribes and lawyers, fully conversant in Jewish law (the *torah*), knew *exactly* what Jesus was doing. Let's take a moment to examine a key Old Testament passage. Recall

> If they thought that Jesus had merely gone off the deep end, the Pharisees could have laughed off his claims.

the story of Moses and the burning bush. He hears a voice from out of the bush, claiming to be "the God of your father, the God of Abraham, the God of Isaac, and the God of Jacob" (Exod. 3:6). Upon hearing that God had chosen him as an instrument to deliver the people of Israel, Moses immediately asks God for a proper name. During

> **Jesus is actually claiming to be YHWH, the one who spoke from out of the burning bush.**

this period of time, Israel was dwelling in Egypt, a land with numerous deities. If Moses was going to make any headway with the Israelites, he needed a name to distinguish this new God from all the rest. God grants his request, and answers, "'I AM WHO I AM'; and He said, 'Thus you shall say to the sons of Israel, "I AM has sent me to you"'" (v. 14). The Hebrew word "I am," which God uses here, is transliterated 'HYH and YHWH is another form of the same verb. Based on a biblical passage (Lev. 24:10ff.) and tradition, Jews by custom refrained from even uttering this sacred name, believing it to be blasphemous to do so.

Now our passages make sense. We understand the Jewish reactions to Jesus; we can understand how they thought Jesus blasphemous. *Jesus is actually claiming to be YHWH, the one who spoke from out of the burning bush*, or, as the Jews put it, a man calling himself God! (8:33). We will examine the entire issue of Jesus' deity later in more detail, but suffice it for now to look at his identity in light of his warning in 8:24. Jesus explicitly links his true identity to whether or not the Jews would die with their sins forgiven. Thus, the question of Jesus' identity takes on new urgency — it truly is part of the "foundation" of Christian theology.

Luke 23:39-43. "And one of the criminals who were hanged there was hurling abuse at him, saying, 'Are you not the Christ? Save yourself and us!' But the other answered, and rebuking him said, 'Do you not even fear God, since you are under the same sentence of condemnation? And we indeed justly, for we are receiving what we deserve for our deeds; but this man has done nothing wrong.' And he was saying, 'Jesus, remember me when you come in your kingdom!' And he said to him, 'Truly I say to you, today you shall be with me in Paradise.'"

Regarding the crucifixion scene, most of the people I talk with are amazed at the faith of the thief on the cross. Considering Jesus' condition, it is incredible that the thief could see in this battered and bloody man any kind of a king, let alone one with a coming kingdom. Although the text is silent on this point, I'm sure that Jesus too was amazed. His words to the thief go beyond his normal reassurances: *here we find the only person in the Bible promised salvation directly — and the promise is from the mouth of Jesus himself!* Such assurance alone is enough reason to examine the thief's confession: what foundational truths are there for us to learn?

First of all, look at what the thief, despite his own terrible agony, was able to grasp:

1) It is right to "fear God," since our deaths do not extinguish His claims upon us (v. 40).

2) Jesus had done "nothing wrong" (v. 41). It is tempting to relegate the thief's admission merely to Jesus' *legal* innocence regarding his being crucified. The thief perhaps was commenting only on the specific injustice of Jesus' execution. But such a reading doesn't do justice to the thief's following words, that Jesus was a king with a kingdom! Mere legal innocence doesn't make one a king! Clearly, the thief links Jesus' "innocence" with his identity as "king," the thief is speaking at some level to Jesus' *character*.

3) Jesus was a king with a coming "kingdom" who could somehow save people who threw themselves on his mercy. It was not amazing that Jesus, a Jewish rabbi, preached about the "Kingdom of God." Such belief and preaching were common in first-century Palestine. What was amazing and entirely new, was that the kingdom was "inextricably bound up with the *person* of Jesus."[10]

> In the thief on the cross we find the only person in the Bible promised salvation directly — from the mouth of Jesus himself!

Had the thief been present earlier in the crowd when Jesus called out (in Luke 12:8), "Everyone who confesses me before men, the Son of Man shall confess him also before the angels of God"? We have no way of knowing. The thief had, however, grasped the essential point that "the verdict passed on men in the final judgment was determined by the attitude they adopted toward Jesus in the present age."[11] **Individual salvation was now tied forevermore to the person of Jesus.** In fact, this focus on the *person* of Jesus is what separates Christianity from every other religion. In Islam, it would be considered blasphemous to make the claims for Mohammed that Christians make for Jesus. In Buddhism, the *teachings* of Buddha are all-important, the following of the "Eightfold Noble Path." Thus, Buddha claimed only to have discovered a way of life, not that he was a "savior" of his followers.[12] As regards Hinduism, nobody knows who the founder even is, let alone tries to worship him! Lastly, Native American religions are primarily ethical religions, with prescriptive teachings sounding very much like those found in Buddhism or Judaism.

> The divine claims of Jesus propel us to take a closer look at the man.

In this chapter, we've looked at three biblical passages, focusing on faith, the "bottom line" of Christianity. Hopefully, they've whetted your appetite to take a deeper look at the identity of Jesus. As we have seen, salvation in Christianity hinges on a belief that's closely related to trust — but not empty trust, nor trust in a mere person, but trust in the Lord Jesus Christ. We have also seen that those who might claim that Jesus never claimed to be anything but a "good person" or a "simple teacher" are simply wrong. In fact, *these claims of Jesus propel us to take a closer look at the man.* Can a divine Jesus be squared with the clear monotheism of the Old Testament? What does the New Testament say about Jesus? Lastly, what is the witness of the early church about Jesus?

---

[1]See the article on belief (*pistis*, "faith") in *The New International Dictionary of New Testament Theology*, Volume 1, Colin Brown, ed. (Grand Rapids: Zondervan, 1975), pp. 593-606.

[2] See the excellent treatment of the usage of "Lord" (*kyrios*) in Brown, *NIDNTT*, Vol. 2, pp. 510-520.

[3] I have not followed tradition in capitalizing pronouns that refer to Jesus (including Scripture quotations) — at least at the first of the volume. I have done this, not from lack of respect, but from not wanting to prejudice the reader before making my case, that is, that Jesus is God. After the case is made, I have switched to the traditional capitalization.

[4] By use of the imperfect tense which shows continuing action.

[5] The pronoun "you" is actually unnecessary here, its usage and location are clearly for emphasis.

[6] The *Encyclopaedia Judaica* lists 18, *The Jewish Encyclopedia* lists 7. *Encyclopaedia Judaica*, Vol. 5, (Jerusalem: Keter Publishing House, 1972), p. 142. *The Jewish Encyclopedia*, Vol. III (New York: KTAV, 1901), p. 554. The apparent discrepancy is due only to the latter's specificity in relegating more crimes to other forms of capital punishment such as burning or "violent means."

[7] See the excellent discussion in Brown, *NIDNTT*, Vol. 3, pp. 340-345.

[8] Ibid., pp. 341-342.

[9] German uses a "j" for the same sound as our English consonant "y." Both represent the same letter in Hebrew, which does not have the sound of our English "j."

[10] Brown, *NIDNTT*, Volume 2, p. 383.

[11] Ibid.

[12] See, for example, *The Encyclopedia of Religion* (Secaucus, NJ: Poplar Books, 1945), p. 94.

REFLECTING ON LESSON ONE

1) Why is a proper foundation important to examining the truth of a religious or philosophical claim? Can you think of other truth claims whose foundations need examining? What is an *objective* truth claim?

2) Why are the passages in this chapter foundational to Christianity? How many others can you think of? If you're in a group setting, this would be an excellent time

to brainstorm, and to come up with as many as you can. Make sure you distinguish between the *cardinal* truths of Christianity, and those peripheral to the faith.

3) In your own words, write down the biblical meaning of "faith."

4) Without looking ahead in this book, can you think of other biblical passages where Jesus claims to be God, or where the writers call him God? See if you can come up with at least five. Is there any particular book of the Bible that seems to have a lot of such claims?

5) Why do you think the one thief on the cross thought that Jesus was a king?

6) Why do you think the Philippian jailer thought that Paul and Silas might have the answer to his question? Does the context of their imprisonment give any clues?

Resources for Further Study

Decide for Yourself: A Theological Workbook by Gordon Lewis. Downers Grove, IL: InterVarsity Press, 1970. An interactive workbook which guides the reader as he or she explores not only all the cardinal doctrines of the faith, but more peripheral ones as well. Examines several theological positions, giving scriptural support for each one.

Essentials in Christian Faith and *Essentials in Christian Practice* by Steve Burris. Joplin, MO: College Press, 1992. An overview of the most important doctrinal issues of Christianity. See also the next listing.

Evangelical Essentials: A Liberal-Evangelical Dialogue by David Edwards and John Stott (Downers Grove, IL: InterVarsity

Press, 1988). Two scholars — one a famous evangelical, the other a more liberal theologian — discuss all the major doctrines of the faith. Notable for its collegial tone, and its depth in looking at the tough issues.

A General Introduction to the Bible by Norm Geisler and William Nix. Chicago: Moody Press, 1981. A nice general survey book that looks at the major issues that surround the writing of the Bible.

Handbook of Basic Bible Texts by John Jefferson Davis. Grand Rapids: Zondervan, 1984. Davis, a leading conservative theologian is quite thorough in his treatment of major Christian doctrines. Like Lewis (above), he examines the differences that exist between different theological schools within Christianity — and the justification for such differences.

Know the Truth by Bruce Milne. Downers Grove, IL: InterVarsity Press, 1982. An excellent handbook of basic Christian doctrines, using historical figures for analysis and comparison.

A Newcomer's Guide to the Bible by Mike Armour. Joplin, MO: College Press, 1999. A survey of the Bible written especially for the seeker who has never read the Bible and knows little or nothing about what it contains.

A Survey of Old Testament Introduction by Gleason Archer. Chicago: Moody Press, 1980. The author is one of the leading evangelical, Old Testament scholars in the world today. In readable fashion, he covers the major questions and issues that surround the Old Testament.

2
TWO

WHO THEN IS THIS?

Now it came about on one of those days, that he and his disciples got into a boat, and he said to them, "Let us go over to the other side of the lake." And they launched out. But as they were sailing along he fell asleep; and a fierce gale of wind descended upon the lake, and they began to be swamped and to be in danger. And they came to him and woke him up, saying, "Master, Master, we are perishing!" And being aroused, he rebuked the wind and the surging waves, and they stopped, and it became calm. And he said to them, "Where is your faith?" And they were fearful and amazed, saying to one another, "Who then is this, that he commands even the winds and the water, and they obey him?" (Luke 8:22-25).

As we have seen, what sets Christianity apart from Judaism and other religions is its crystal clear focus upon the *person* of Jesus. Although Christianity has something to say about the *behavior* of those who call themselves Christians, it is not primarily an ethical religion, as is Confucianism or Buddhism. Most of the ethical precepts found in the New Testament can be found in many religious traditions. The so-called "Golden Rule," for example, is found in traditions as various as Rabbinic Judaism, Native American religion, Buddhism, and Confucianism. Likewise, it is wrong to reduce Jesus to the level of a moral teacher. The claims of Jesus himself, as well as the writings of those who knew him, simply don't paint the picture of a moral guru who came to deliver his own special brand of teachings. His sayings regarding his identity and mission are far too audacious; and the claims of others go far beyond such a portrait. In fact, as C.S. Lewis has pointed out, the words of Jesus paint the options quite starkly: "*The idea of a great moral teacher saying what Christ said is out of the question. In my opinion, the only person who can say that sort of thing is either God or a complete lunatic suffering from that form of delusion which undermines the whole mind of man.*"[1]

As we shall see, the question of the identity of Jesus is perhaps the greatest difference between Christianity and Jehovah's Witnesses. Not only because of the great mistake in his *identity*, but in what such a mistake means in terms of salvation. Let's take a closer look at the person of Jesus, focusing on three sources: the Old Testament, the New Testament, and the early Church.

THE OLD TESTAMENT

Judaism is, and has always been, a monotheistic (one God) religion. That is, the creed, "Hear O Israel! The LORD is our God, the LORD is one!" (Deut. 6:4) has always been part and parcel of what it means to be a practicing Jew. In stark contrast to the polytheistic (many Gods) religions that inhabited the land of Canaan before the Israelites entered, God developed a special covenant relationship with Israel that had as its first and foremost requirement an acknowledgment that there is only one God (Exod. 20:2-4). Such an acknowledgement separated the Jews from their surrounding neighbors, and became the defining characteristic of what it meant to be a Jew. *It was part of the high calling of the prophets to remind Israel of this heritage,* and any lapse into polytheistic worship was a sure sign of a weakening relationship between the Lord and his people (Isa. 43:10; 44:6,8).

> **The idea of a merely human great moral teacher saying what Christ said is out of the question.**

With all these warnings and admonitions in mind, imagine yourself a first-century Jew. You have been taught since birth that there is only one God and Lord; in all likelihood, Deuteronomy 6:4 has been written on the doorposts of your home, to be reverently touched upon going and leaving. Being a monotheist is in your blood. You understand well that your nation's special relationship with *Yahweh* depends on your proper worship of this one God.

> It was part of the high calling of the prophets to remind Israel of their heritage of montheism.

Then you hear of strange happenings in Galilee. There is a man who heals the sick, performs miracles of all kinds, and makes startling personal claims. Is it no wonder that it would be difficult — perhaps nearly impossible — to accept that this *man* was God? And, even if you happened to be one of this man's closest friends, is it not reasonable that you would still have reservations about calling him God? However, if we take a moment to examine the monotheistic fabric of the Old Testament, we can see where Christ's deity is not in conflict with God's unity.

First of all, consider the Hebrew word for one, the same word used to tell the Jews that there is "one" God, *echad*. The word can carry with it the idea of plurality, or better, the notion of plural unity. The same word is used in Genesis (2:24) when God says that husband and wife will become "one" flesh in their union. Surely it is obvious that husbands and wives retain their individual bodies when married; the oneness spoken of is a plural unity, two persons, yet one in purpose, in union.

Such a plural unity has other support in the Old Testament. When God creates humanity, his words are "Let *Us* make man in *Our* image" (Gen. 1:26, emphasis mine). The plural pronouns allow for a divine plural unity within the Godhead. There is certainly no textual or theological support to suggest that the "our" has anything to do with angels as cocreators, or that human beings are made in the image of angels. Clearly, the divine image that is humanity's blueprint is from God and God alone.

Likewise, when Isaiah receives his prophetic commission from God, he is responding to God's question, "Whom shall I send, and who will go for *Us*?" (Isa. 6:8, emphasis mine). Again, the plural pronoun carries with it divine sanction and weight: God is doing the sending, the commissioning. God is doing the talking, and Scripture nowhere suggests that angels possess the authority to commission prophets.

Thus, while it is certainly true that the Old Testament may not explicitly teach the doctrine of the Trinity (which we'll cover later), it most certainly allows for it. This is especially the case when one looks at some of the prophetic passages that refer to the coming Messiah.

Every Christmas season, church choirs everywhere prepare to sing Handel's *Messiah*. One of the piece's most famous parts includes the words of the prophet Isaiah, speaking of the future Messiah: "For unto us a child is born, unto us a son is given: and the government shall be upon his shoulders: and his name shall be called Wonderful, Counsellor, The mighty God, The everlasting Father, The Prince of Peace" (Isa. 9:6, KJV). *This passage clearly states that the Messiah's name is the "Mighty God," the "Eternal Father."* What other Father or God could Isaiah possibly have in mind except the God of Abraham, Isaac, and Jacob? Moses himself writes that Israel's God is the "mighty and awesome God" (Deut. 10:17). Thus, when Isaiah, a Jew steeped in the monotheistic tradition of Judaism, refers to the future Messiah as "Mighty God," it is clearly the God of his forefathers.

There is also another way to examine how the Old Testament writers looked at the identity of the coming Messiah. It has to do with a device called a *deductive argument*. Let's use a modern example. If a company brochure states that the President of the company is the only one who may be called "Fearless Leader" in company circles, and then you read "Fearless Leader" on the door

> **Isaiah clearly states that the Messiah's name is the "Mighty God," the "Eternal Father."**

of your friend Ralph, then it's safe to assume that Ralph is President. Or, stated another way:

1) **"Fearless Leader" is a title reserved exclusively for the company president.**

2) **Ralph has the title "Fearless Leader."**

3) **Thus, Ralph is president.**

> Writers of Scripture knew that the real king of Israel was God Himself.

This sort of reasoning is fairly common in the Old Testament, when prophets wrote of the coming Messiah. Take for example the idea of king. Who is the real king of Israel? Granted, the country had its share of human kings, most notably Saul, David, and Solomon. But the prophets and *writers of Scripture knew that the real king of Israel was God Himself.* David, in fact, had no qualms about calling God his king (Ps. 5:2); and it was "the Lord" who was "king forever and ever" (Ps. 10:16). Isaiah, speaking for the Lord, says "I am the LORD, your Holy One, the Creator of Israel, your King" (43:15); and later on, Isaiah calls the Lord "the King of Israel" (44:6).

With this idea of kingship in mind, take a look at what the prophet Zechariah says about the Messiah. In a passage universally acknowledged as messianic, Zechariah announces "Behold your king is coming to you; He is just and endowed with salvation, Humble, and mounted on a donkey, Even on a colt, the foal of a donkey" (9:9). It is clear that Zechariah had in mind the ultimate king of Israel when he penned this passage. None of the human kings of Israel were ever described as being "endowed with salvation," and thus the passage clearly refers to God as both the king and savior of Israel. And who is it that rode into Jerusalem on a donkey, who rode in accompanied by the cheers of the crowds? Clearly, Jesus fulfilled this prophecy when he rode into Jerusalem the week before his crucifixion (Matt. 21:5). Thus, the analogy is complete, and our deductive argument is as follows:

1) **God is the ultimate king of Israel.**

2) **Zechariah called the Messiah the ultimate king of Israel.**

3) **Jesus was the Messiah.**

4) **Thus, Jesus is the ultimate king, and hence God, of Israel.**

Thus, when the title "King of the Jews" was affixed on the cross of Jesus, although the Roman authorities intended the title as a mocking joke, the words spelled out more than some satirical jest. Jesus, as God, is indeed the King of the Jews.

Using the same sort of reasoning, let's take a brief look at the concept of *savior*. Any believing Jew knew full well that there was only one savior of Israel, God. Take a moment to open your Bibles to the book of Isaiah. Here, the prophet takes great pains to state that God is the savior of Israel: "For I am the LORD your God, the Holy One of Israel, your Savior" (43:3). The Lord even goes so far as to say that "there is no savior besides Me" (43:11). The prophet Hosea echoes Isaiah when he states "Yet I have been the LORD your God Since the land of Egypt; And you were not to know any god except Me, For there is no savior besides Me" (13:4). The Psalmist states that Israel's deliverance from Egypt was quickly forgotten by the ungrateful people because they had forgotten "God their Savior" (Ps. 106:21). *These passages provide a very real dilemma for those who want to call Jesus "savior" without acknowledging that he is God.* Clearly, God reserves the title "savior" for Himself alone. Yet, just as clearly, the New Testament shouts with one voice that Jesus is the savior of not only Israel, but the whole world (1 John 4:14). Thus, our deductive argument would look like this:

1) **In the Old Testament, God reserves the title "savior" for Himself alone.**

2) **Jesus is commonly called "savior" in the New Testament.**

3) **Thus, Jesus is equated with God.**

> There is a very real dilemma for those who want to call Jesus "savior" without acknowledging that he is God.

Again, this examination of the Old Testament is not intended to be exhaustive. It does, however, illustrate the fact that the Old Testament allows for the idea that God might be a plural unity. It also speaks to the divine nature of the coming messiah. And finally, it puts to rest the idea that the Old Testament is necessarily at odds with the notion that God is a trinity. Like looking at a flower just starting to bloom, with the shades of color just barely noticeable behind the petals, *the Old Testament bears witness to a coming fuller revelation of God and the messiah.* The New Testament is that fuller revelation.

> The Old Testament bears witness to a coming fuller revelation of God and the messiah.

THE NEW TESTAMENT

Nearly all the authors of the New Testament were converts from Judaism, having been steeped in the monotheistic traditions of that faith since their youths. Thus, when they speak of the deity of Jesus, it must not be taken lightly that these men were willing to identify the man Jesus with the God of Abraham, Isaac, and Jacob. Thus, it is amazing that there exist so many direct and clear references to Christ's deity. We have already dealt with Jesus' amazing claim in John 8, but let's look at more references.

Direct References. Keep in mind that this list is not meant to be exhaustive, but thorough enough to prove how consistently the Bible sees Jesus as God.

Thomas answered and said to him, "My Lord and my God!" (John 20:28). Many of us have heard the story of how Thomas, who wasn't with the disciples when Jesus first appeared to them after the resurrection, needed to see "for himself" that Jesus was indeed alive. Indeed, still today, when someone is called a "doubting Thomas," the title infers a person who needs cold, hard, facts in order to believe something. When Jesus appeared to the disciples a second time, he gave Thomas the opportunity he had asked for — to actually touch the wounds Jesus had received upon the cross. When the risen Lord appears, Thomas forgoes the opportunity for "proof," and exclaims that indeed, Jesus is God! If Thomas had been wrong, this would have been the perfect opportunity for Jesus to correct him — after all, as any Jew knew, to label Jesus falsely as God was to blaspheme. Instead, however, *Jesus accepts Thomas's declaration, and pronounces a blessing on those believers who would echo Thomas's words* without the need for a physical demonstration (v. 29).

By the righteousness of our God and Savior, Jesus Christ (2 Peter 1:1). Grammatically, the meaning of the sentence is clear: "God" and "Savior" refer to the same person, Jesus Christ. It reads identically to the English construction: "I wish to thank our boss and good friend, Mrs. Jones," wherein "boss" and "good friend" refer to exactly the same person. The Apostle Peter left no doubt that he intended to call Jesus God; had he intended otherwise, he would have used an entirely different sentence construction.

The church of God, which He purchased with His own blood (Acts 20:28). The meaning here is clear: the pronouns "He" and "His" refer back to God, meaning that the one who bought (redeemed) the church is God. Since Jesus is obviously in mind as the redeemer, then Luke clearly identifies him with being God. Although it is true that some ancient manuscripts substitute "the Lord" for "God," nevertheless, the better reading is to retain "God." The fact that some manuscripts contain "the Lord" no doubt reflects a reluctance by some early Jewish-Christian copyists to ascribe so blatantly physical characteristics (as having blood) to a God who was on record as being spirit.

> Jesus accepts Thomas's declaration, and pronounces a blessing on those believers who would echo Thomas's words.

The appearing of the glory of our great God and Savior, Christ Jesus (Titus 2:13). Here again, the meaning is unambiguous: the titles "God" and "Savior" both refer to Jesus, clearly spelling out his deity. The Greek text does make another rendering possible, but it clearly maintains the thrust of the first: "The appearing of the glory of the great God and our Savior, Christ Jesus." That there is only one person in mind, not two, is made plain by the context which goes on to say that this "great God and Savior" "gave Himself (singular) for us, that He (singular) might redeem us" (v. 14). Thus, Jesus is

> The Gospel of John presents the deity of Jesus in its clearest form.

clearly called both God and Savior, and his role as redeemer in confirmed.

No man has seen God at any time; the only begotten God, who is in the bosom of the Father, He has explained *Him* (John 1:18). More so than any other New Testament book, *the Gospel of John presents the deity of Jesus in its clearest form.* Again and again, John shares with his readers his passion that this Jesus, whom he had known well for some three years, is indeed the God who created and sustains the universe. In this passage, John emphasizes Jesus as the *revealing* God. That is, just as it is foolish to ask an ant to explain a human, so also only God can *explain* God. Here, Jesus is uniquely qualified to explain God the Father, due to his own status as God the Son. A few manuscripts read "the only begotten Son," but by far the best textual tradition retains "the only begotten God."

But of the Son, *He says*, "THY THRONE, O GOD, IS FOREVER AND EVER" (Heb. 1:8). In this undisputed text, the author of Hebrews states unequivocally that the Son is God. Take a moment to look at the context of the above passage. The entire context of Hebrews is to elevate the mission and identity of Jesus — to say how much superior is his kingdom and reign than the Old Covenant types. The writer here is quoting the Psalms (45:6), and the song found there of adoration and praise to God is taken by the writer of Hebrews and attributed to the Son (Jesus).

In the beginning was the Word, and the Word was with God, and the Word was God (John 1:1). Since John writes later (v. 14) of the Word becoming flesh, we know that when he equates the Word with God, that he is equating Jesus with God. Again we see John's insistence on the deity of Jesus —

truly this gospel is a monument to the astounding truth that God has indeed walked with us (which, after all, is the meaning of *Emmanuel* in Isa. 7:14).

Perhaps no other verse of the Bible has been such a battleground between Jehovah's Witnesses and Christians as this one. Entire articles have been written concerning just this one verse and its correct translation. I will have more to say on this verse later. For now, let me say this: as one who has taught Greek for many years, I would not pass a first year Greek student if he or she could not translate John 1:1 correctly. And any correct rendering of the Greek original must be faithful to John's absolute insistence on the identity of the Word being God.

For in him [Jesus] all the fullness of Deity dwells in bodily form (Col. 2:9). In our last passage, Paul states that all of "what makes God, God" is in Jesus. What then does God the Father possess that Jesus lacks? Nothing! All means all. Thus, Paul's claim here echoes the words of Jesus to Philip: "He who has seen Me has seen the Father" (John 14:9).

Deductive Arguments. Just as the case with such arguments which had their basis in the Old Testament, so also there are similar arguments in the New Testament. In these cases, *Jesus claims divine prerogatives reserved for God alone.*

    1) **Jesus as the object of worship.** Remember our discussion on blasphemy? Blasphemy means trying either to bring God down to man's level, or to take man up to God's. In either case, it was an extremely serious offense in the Jewish world, one of those which merited death. One of the most blatant forms of blasphemy was to worship any other being except God — neither men nor angels, but only God may be worshiped (Acts 10:26; 14:15; Rev. 19:10; 22:9). However, Matthew doesn't hesitate to mention that as Jesus ascended into heaven, people worshiped him (Matt. 28:17). Clearly, the context indicates that the object of worship was Jesus himself, not God in general. Otherwise, when

> Perhaps no other verse of the Bible has been such a battleground between Jehovah's Witnesses and Christians as John 1:1.

the text mentions that "some doubted" in contrast to the worshiping believers, who else could be the object of doubt than Jesus himself? Jesus is again worshiped when the twenty-four elders "fall down before the Lamb" (Rev. 5:8). The Lamb, of course, is none other

> Jesus claims divine prerogatives reserved for God alone.

than Jesus himself, the one who was "slain" (v. 9), who "purchased people from every tribe and tongue and people and nation." Thus, again we have a clear deductive argument for the deity of Jesus: A) Only God may be worshiped; B) People worship Jesus; therefore, C) Jesus must be God.

2) **Jesus is the object of prayer**. Nothing could be more obvious than the fact that prayer should be directed toward God. We are nowhere instructed to pray either to angels or people. But look at Acts 7:59. Here Stephen, the first Christian martyr, prays to Jesus to "receive his spirit." The language of prayer is instructive. Anyplace in the Scriptures where one reads that people "called upon the Name" of God is equivalent to "praying to God" (compare Joel 2:32 with Acts 10:12). With that in mind, examine the following passages: Acts 9:14,21; Romans 10:12; and 1 Corinthians 1:2. What do they all have in common? Calling upon the name of Jesus (that is, praying to Jesus) is equivalent to calling upon the name of the Lord (of the Old Testament). Take a moment to look at Romans 10:12-13. Here, Paul explicitly calls for prayer to Jesus for one's salvation, then calls such prayer praying to *Yahweh* Himself, quoting again the prophet Joel (v. 13). Thus, our deductive argument: A) We may pray only to God; B) Prayer to Jesus is encouraged throughout the New Testament; thus, C) Jesus is God.

3) **Jesus was able to forgive sins**. At first glance, forgiving sins seems to be the right of every Christian — what's so special about Jesus doing it? The difference, however, lies in just whose sins we are forgiving. Paul clearly

states that we are to forgive one another for slights or wrongs committed against us (Eph. 4:32). Likewise, in the so-called "Lord's Prayer," Jesus instructed all believers to ask God to forgive them their debts as we forgive those who have slighted us (Matt. 6:12). But, it is entirely another matter altogether to grant forgiveness to a third party not directly connected to you. Such is the scene when Jesus heals the paralytic who was lowered to him through the roof (Mark 2:1-12). Here, Jesus tells the paralytic (the third party) that his sins (against others) are forgiven (v. 5). Not all the audience were joyful, however. The scribes, who knew well just who had authority to forgive such sins, thought to themselves, this man is blaspheming: "Who can forgive sins but God alone?" (v. 7). Jesus knew exactly what they were thinking (yet another argument for his deity), and gave the crowd a physical sign (healing the paralytic) to demonstrate (vv. 10-11) that he had authority to forgive sins (a spiritual, nonvisible power). He never disputes the scribes' reasoning that "God alone" can forgive sins; instead, in a roundabout way, he makes an astonishing claim for his identity. With the information we've learned about Jesus' forgiving sins, and the two models above, can you give the deductive argument here for Christ's deity, using his ability to forgive sins?

We've now looked at a representative witness of the biblical materials. *There is a consistent and persistent voice to Jesus' deity found in Scripture.* Believing that history is an important and valuable record for our topic, let's take a brief look at what the early church said about the identity of Jesus.

THE WITNESS OF THE EARLY CHURCH

> There is a consistent and persistent voice to Jesus' deity found in Scripture.

During the first three centuries or so of the church's existence, it had a rocky relationship with Rome, the empire that stretched its control over much of the known world. Not until the Edict of

Milan in 313 and the subsequent conversion of the Roman Emperor Constantine did official persecutions of the church finally cease. Thus, the earliest church fathers were far too preoccupied with staying alive to worry about writing huge tomes regarding Christian theology. Nevertheless, *there is a consistent witness to Christ's deity from the "Apostolic Fathers"* (the generation immediately following the apostles) to the "Apologists" (those Christian thinkers from 120-220) finally to those Fathers who immediately preceded Constantine's conversion (late third to early fourth century).[3]

> There is an equally consistent witness to Christ's deity from the "Apostolic Fathers" on.

The Apostolic Fathers. Polycarp (ca. 70-156), a disciple of the Apostle John, recited a common prayer immediately before his martyrdom. Its formula is what modern theologians would call "trinitarian," with the Father, Son, and Holy Spirit all mentioned and given equal stature.

Clement of Rome (who wrote around 90-100) may be the Clement mentioned in Paul's letter to the Philippians (4:3). In any case, Clement clearly takes Christ's preexistence for granted, saying "it was he who spoke through the Spirit in the Psalms."[4] In addition, the letter *2 Clement* (attributed to Clement) opens by admonishing the readers to "think of Jesus Christ as of God."[5]

Ignatius of Antioch (who was martyred in 117) declares that Jesus "is our God" describing him as "God incarnate," and "God made manifest as man."[6] His theology and terminology parallels the Scriptures' at key points, and does not hesitate to speak of Jesus' "divine Sonship."

The Apologists. As the name indicates, these Christian thinkers were known for *defending* (the Greek *apologia* means "a defense") Christianity against pagan and competing philosophies. Thus, a new more intellectual formula and context are introduced. Justin, sometimes called Justin Martyr (ca. 100-165), echoed the Apostle John when he called

Jesus the *logos* (the word, as found in John 1:1). Justin went on to state that this divine Word indeed "is God."[7]

Tatian (ca. 110-172), Theophilus of Antioch (ca. 175), and Athenagoras (ca. 2nd century) all stressed the divine nature of the *logos* or Word. True enough, these apologists never developed a full-blown doctrine of the Trinity, but their understanding never wavered that Jesus as the Word "was one in essence with the Father."[8]

Irenaeus. One of the most dominant Christian thinkers of this important period was Irenaeus (who wrote from ca. 175-195). As a boy he had sat at the feet of Polycarp, and his thought and influence was pervasive up to and including the time of the first ecumenical church council at Nicea (in 325). Influenced by the apologists, he was less philosophical than they and was more self-consciously theological. Irenaeus has perhaps the fullest and most complete doctrine of the Trinity of all the early church fathers. *He insisted that the Son is fully divine, for "whatever is begotten of God is God."*[9] Thus, there exists from the New Testament documents all the way to Nicea an unbroken stream of tradition that agrees fully with biblical insistence that Jesus is God.

WHAT CAN WE LEARN FROM EARLY CHURCH CREEDS?

This biblical/theological tradition found its way into all the various ecumenical creeds of the early church. As Philip Schaff, the famous historian has noted, creeds act as summaries of biblical doctrines and can serve as guards to false doctrines and practices.[10] Since, by their very name, "ecumenical" creeds signify doctrine agreed upon by all Christian churches, they are instructive as to their picture of Jesus. Let's examine briefly these four ecumenical creeds.

> Irenaeus insisted that the Son is fully divine, for "whatever is begotten of God is God."

The Apostles' Creed (ca. 200). *Although it is not, as might be thought, from the pens of the apostles, this creed nevertheless* *encapsulates apostolic teaching.* Scholars agree that the creed was never meant to be a full exposition of Christian teaching but a summary of doctrine, perhaps to be recited by a newly baptized convert. Thus, further development of its tenets was inevitable — in many instances, its words depend on a community already in agreement about their meaning. As the church grew and expanded, it would become necessary to elaborate just what was meant by some of the key words and phrases.

> Although not from the pens of the apostles, the "Apostles' Creed" encapsulates apostolic teaching.

Regarding the person of Jesus, the text of the creed speaks of Jesus as the "only begotten Son our Lord; who was conceived by the Holy Spirit, born of the Virgin Mary; suffered under Pontius Pilate, was crucified, dead, and buried; he descended into hell; the third day he rose from the dead; he ascended into heaven; and sitteth at the right hand of God the Father Almighty; from there he shall come to judge the living and the dead." When one considers the theological environment from which this creed came, it is clear that when it calls Jesus "the only begotten Son," that such is a reference to his deity. As Irenaeus, a contemporary of the Creed, had said, whatever is begotten of God is God. Nevertheless, as the church faced challenges from within, it became necessary to spell out more clearly just what was orthodoxy and what was heresy.

The Nicene Creed (325). The first attempt at a comprehensive creed that was both thorough *and* specific in its use of theological language is the Nicene Creed. A product of the Council of Nicea, it was expressly designed to combat Arianism, a belief that Jesus was not fully God and hence was not equal to the Father. What follows is the relevant portion of the Creed, that part which expands on the Apostles' Creed of the previous century. The Creed states that "We believe . . . in one Lord Jesus Christ, the Son of God, begotten of the Father, Light of Light, very God of very God, begotten not made, being of one substance with the Father; by whom all things were made; who for us men, and for our salvation, came down and was incarnate and was made man."[11] The Creed takes great pains to explain just what "begotten" means — that Jesus is in no way "created," and that he is identical "in substance" to the Father.

This view that Jesus was identical to the Father contrasted sharply with the views held by a certain Bishop Arius of Alexandria. Arius claimed that "there was a time when the Son was not," effectively denying the eternal existence of the Word. In this belief, he was similar to the Ebionites, a small Jewish-Christian sect that denied both the virgin birth and deity of Jesus. At the Council Arius was confronted by the writings of a young deacon named Athanasius who insisted, as both the Apostle John and Irenaeus had before him, that the Word (Jesus) was fully God. Thanks to the influence of Athanasius, Arianism was roundly condemned at the Council, and *the deity of Christ was never seriously challenged in the church again.*

The Athanasian Creed (ca. 400). This creed represents a substantial expansion of the Nicene Creed (which itself was an expansion of the Apostles' Creed). The deity of Christ is upheld, and it expands even further the ramifications of Christ's equality with the Father. "And the universal faith is this: That we worship one God in Trinity, and Trinity in Unity . . . So the Father is God; the Son is God; and the Holy Spirit is God."[12]

Here for the first time in a creed is the word "Trinity" (although the concept is certainly biblical). Utilizing Scripture and creeds thus far, we might define the Trinity this way: **One God, eternally existent in three persons, the Father, the Son, and the Holy Spirit. All are codivine, coequal, and coeternal.**

> The deity of Christ was never seriously challenged in the church again after Nicea.

The Chalcedon Creed (451). This creed was an attempt to probe the person of Christ, especially to examine how he relates to mankind. For if Jesus were truly God because of the apostolic witness, he surely was just as truly man for the same reason. Didn't the apostles record Jesus eating, tiring, weeping?

Since Jesus is *humanity's* savior, must he not be *human*? "We confess our Lord Jesus Christ, the same perfect in Godhead and also perfect in manhood; truly God and truly man, of a rational soul and body; consubstantial with the Father according to the Godhead, and consubstantial with us according to the Manhood; in all things like us, without sin."

> Witness to the deity of Christ is anchored by Scripture, witnessed and defended by the Church Fathers, and explained in the ecumenical creeds.

Thus, this creed — although more specific and technical than the preceding two — contains no new or startling dogma. It is considered the last of the ecumenical creeds, those creeds that are believed by all Christians. These creeds serve two important functions: 1) to help Christians understand there's a long and consistent *history* to their faith; and 2) to set boundaries between those of the faith and those outside.

As regards the former function, it should be clear to the reader that there is *a consistent witness to the deity of Christ that is anchored by Scripture, witnessed and defended by the Church Fathers, and explained in the ecumenical creeds*. These creeds are nourished in the soil of the Bible. In this very important sense, no Christian is alone. She or he is joined to the fellowship of those who have paved the way, who have thought through these tough issues.

The ecumenical creeds, faithful to biblical teaching as they are, also serve as fences that mark off Christianity from heresy. Nobody of good will wishes to be judgmental, but the simple fact of the matter is that these creeds help us understand that not everybody is a Christian who claims the title. Christianity has always been known as an "exclusive religion," that is, one that believes that some people are Christian, others aren't. The "bottom line" to Christianity, of course, must be gleaned from Scripture alone — salvation doesn't ultimately depend on adherence to any creed. Creeds, unlike Scripture, are not God's Word. But, as we have seen, the ecumenical creeds speak with the same voice as Scripture to the amazing fact that, some two thousand years ago, the very same God who created the universe by a word, walked, talked, and laughed among us.

---

[1] C.S. Lewis, *God in the Dock* (Grand Rapids: Erdmanns, 1970), p. 158.

[2] I am indebted for much of the following material from the classic *Early Christian Doctrines* by J.N.D. Kelly (Edinburgh: R. and R. Clark, 1960).

[3] See Kelly, *Early*, p. 91.

[4] Ibid.

[5] Ibid., p. 92.

[6] Ibid., p. 97.

[7] Ibid., p. 101.

[8] Ibid., p. 107.

[9] See Schaff's monumental 3-volume work, *The Creeds of Christendom* (Grand Rapids: Baker, 1993), Volume I, p. 8.

[10] This is the original Creed of AD 325, found in Schaff, *Creeds*, Vol. I, p. 29.

[11] Schaff, *Creeds*, Vol. II, pp. 66-67.

REFLECTING ON LESSON TWO

1. How can the monotheistic Old Testament allow for Jesus being God?

2. Using either the Old or New Testament materials, construct a "deductive argument" for the deity of Jesus. If you can construct five such arguments, pat yourself on the back! If you can think of three, silently rejoice. Two or less, reread the part of the chapter that deals with deductive arguments.

3. Try to come up with four direct references to the deity of Christ. What New Testament book does the author think is particularly helpful in this regard? How might this

knowledge help when talking about Jesus to real-life Jehovah's Witnesses?

4. Why are the early church fathers important when examining the person of Jesus?

5. What is an "Ecumenical Creed?" Why are they helpful, according to the author?

RESOURCES FOR FURTHER STUDY

The New International Dictionary of the Christian Church by J.D. Douglas (ed.). Grand Rapids: Zondervan, 1978 edition. This is a wonderful work for studying virtually any topic in religion that touches upon Christian theology. Arranged topically with helpful bibliographies.

The Creeds of Christendom by Philip Schaff (bibliographic information in the notes). A nice scholarly set for the serious student of history. Copious notes so the reader can, if he wishes, dig even deeper.

Early Christian Doctrines by J.N.D. Kelly (bibliographic information in the notes). A great, thorough work that takes the reader from the Apostolic Age to the Middle ages. Not for the beginner.

A History of the Church: From Pentecost to Present by James North. (Joplin, MO: College Press, 1983). An excellent overview of the developments in movements influencing church doctrine and organization for the past 2,000 years.

THREE

SALVATION, RESURRECTION, AND THE SECOND COMING

This is the third and last chapter in the "toolbox" section of the book. It covers three more essential doctrines of the Christian faith — and three doctrines for which the Jehovah's Witnesses impose radically different interpretations. In this chapter, we're covering the concepts of salvation and resurrection, and the return of Christ to earth.

I know I know. I can just hear you saying "Wait a minute, McKenzie! I can buy that both the deity of Christ and salvation by grace are essential doctrines, but the *resurrection*? The *second coming*? How are these foundational?"

Granted, it is entirely *possible* to be a Christian with a false view of the resurrection, or even to have a misguided notion on just how Christ is coming again. It is doubtful, for example, if the thief on the cross had all his eschatological ducks in a row, so to speak (But even the thief knew Jesus was coming with a kingdom!).

However, two further replies are in order. First, keep in mind that all doctrines are related. That is, like dominos standing on edge in a line, when you adjust or alter one doctrine you create results in others. To use just one example, if one denies the bodily resurrection of Christ (as Jehovah's Witnesses do), then such a denial will undoubtedly affect how a person views His second coming (as it has with the Witnesses). If Jesus doesn't have a physical body, then in what form will He come again? You get the picture. Yes, it's possible to be a believer with an inconsistent theology, but *an inconsistent theology is an unstable theology.* Like

> An inconsistent theology is an unstable theology.

the second law of thermodynamics in the physical realm, such a belief system tends to break down into disorder. Such is the point of *Why God Became Man* by Anselm of Canterbury (d. 1109). Anselm's great contribution to Christian theology was that Jesus *had* to be both fully God and fully man: divine, because the death of one mere human could not save humanity; human, because only then could a saving God totally identify with us. Thus, a proper understanding of the person of Christ leads nicely to a correct understanding of salvation.

The second reason for addressing these doctrines relates to the first. To deny these important doctrines of the Christian faith is to place oneself on a very dangerous and slippery slope. If one denies, for example, that salvation is by grace (a doctrine taught so clearly in Scripture), then on what basis does he or she affirm the other doctrines of the faith? Thus, there is ample reason to place all this chapter's doctrines in the "essential" category.

SALVATION

Most all world religions have some concept of "salvation." They may *mean* different things when they use the word, but, in general, salvation is a concept familiar to many religious traditions. Christianity is certainly no different. In fact, one may argue that salvation is probably the most important theme of the Bible!

I recall one of my favorite television shows, *Frasier*. In this particular episode, the housekeeper Daphne Moon gives Frasier's dad Martin a beautiful new sweater for no particular reason. Martin is thankful but flabbergasted. "This is really great, but what did you get me this for? It's not my birthday or anything!" Daphne replies, "I know, I just wanted to get you something." Martin's final reply is all too familiar: "Oh . . . now I'll have to get you something."

Martin's answers to Daphne are both funny and tragic. I suspect both reactions stem from the same source: it's a purely human response to want to "pay somebody back" for a good deed, to "respond in kind" to a good turn. That's simply in our nature. The tragedy comes into play when somebody just wants to do something nice for someone — without getting something in return. Then, all too often, we scurry about to pay them back, not letting the other's gift just rest there as an uplift to our soul. We somehow want to "earn" the other's gift. In some rural areas of our country, it's sometimes put this way: "We don't want to be beholdin' to anybody."

Theologically, this human tendency pops up again and again. Nothing can be so clear as the great truth — nearly lost at times — that Christian salvation is an unearned gift from the Lord. There is nothing, absolutely nothing, we can do to earn or merit this gift. *This is the great truth of Christianity, to be safeguarded at all costs.* Let's take a look at the New Testament to see how Scripture views salvation. I have divided the concept of salvation into three parts, with each contributing a vital link toward understanding the whole.

The Law. What part can the law have in our salvation, especially considering our salvation is a gift? Good question. The apostolic writers considered the law as both a divine reminder and a signpost, never a code to be followed in order to earn salvation. It serves as a reminder of our failure to live rightly, to perform its demands; and it also serves as a signpost to the gospel, which is the example *par excellence* of God's grace and mercy.

Through the Law *comes* the knowledge of sin (Rom. 3:20).

> Grace is the great truth of Christianity, to be safeguarded at all costs.

The Law brings about wrath, but where there is no law, neither is there violation (Rom. 4:15).

Sin is not imputed when there is no law (Rom. 5:13).

The Law came in that the transgression might increase (Rom. 5:20).

I would not have come to know sin except through the Law (Rom. 7:7).

> The Law was designed to act as a spur to help us find the cross of Christ.

Thus, Paul drives home the point that the Law brings knowledge, and it is not good news! Oh yes, and for those who think they've escaped the Law's demands because they've never read the 10 commandments, Paul closes that loophole by affirming that *all* people have the Law (indeed, the 10 commandments) written in their hearts (Rom. 2:14-15). This ought to be the death knell for all those who desire righteousness by following moral codes — such is a fruitless endeavor. But, besides bringing the awful news of our human failures, the law serves *to point toward* the good news.

The Law has become our tutor *to lead us* to Christ, that we may be justified by faith (Gal. 3:24).

In this very interesting passage, Paul calls the Law a tutor, in Greek a *paidagogos*. During Paul's time, a tutor was one (usually a slave) whose responsibility it was to superintend the conduct of the young boys and girls in his charge. His duties were not the positive ones of the teacher (do this and that), but rather those more of a truant officer (don't do this and that).[1] In other words, the tutor's job was to tell his charges what *not* to do. Thus, Paul's point about the Law becomes clear. Its primary task was not to tell us what to do (and certainly not how to attain righteousness), but, by its constant negative prohibitions, *to act as a spur to help us find the cross of Christ*.

Perhaps the best way to see this function of the Law is by the following story. A wealthy philanthropist named Howard died and was immediately ushered into the presence of St. Peter. After a short period of mourning, Howard became happy, contemplating the joys that must surely await him. St. Peter finished some paperwork, then eyed the newcomer over. "Greetings, friend! Welcome to the portals of heaven. I will ask you a series of questions, and, depending on your answers, I will reward you with a certain number of points. You will need one thousand points to get into heaven."

Despite the seemingly large number, Howard was not worried. He knew he had led a good life. "Ask away, my good man."

"Very well. Did you ever cheat on your taxes?" asked Peter.

"Certainly not!" Howard replied, somewhat offended.

"Excellent," retorted Peter. "We don't get many like you. One point. Were you ever unfaithful to your wife?"

Somewhat stunned by only receiving one point for his lifetime of honesty, Howard mopped the sweat off his brow as he gathered himself for the next answer. "Of course not!" he flashed. "I loved my wife my whole life!"

"This is truly marvelous. You're one in a million! You get another point," answered Peter, as he dutifully recorded the tally. "Now, just a couple more questions — "

"A couple more? What kind of place is this?" exploded Howard. "But *by the grace of God* nobody would get in — "

"You've just earned your thousand points," Peter breezed. "Go on in."

Like Howard, almost pushed into grace by a true look at his own futile attempts at righteousness, we need to see that the Law is God's way of telling us He doesn't grade on the curve.

Justification. *Justification is the act of God that declares us righteous.* In human timing, it is the moment we are "saved," the instant we are looked upon as possessing the perfect righteousness of Christ. As we now know, such righteousness does not — indeed, cannot — come from following the law.

> **Justification is the act of God that declares us righteous.**

For we maintain that a man is justified by faith apart from works of the Law (Rom. 3:28).

We have believed in Christ Jesus, that we may be justified by faith in Christ,

and not by the works of the Law; since by the works of the Law shall no flesh be justified (Gal. 2:16).

> One's faith is powerful — if placed in Jesus.

Here, the common thread is a realization that human beings are saved apart from trying to follow the 10 commandments. Following the Law is thus juxtaposed with personal faith in the person of Christ. The Law is powerless to save; *one's faith is powerful — if placed in Jesus.*

Grace and Faith. It is best to see these two parts of salvation as two sides to the same heavenly coin. We did not earn this coin, it indeed is a gift from God. Yet, one side (faith) is meant to have a very human component — it is meant to be seen, and to result in good works. Not of course to "earn" this coin, but as a response of gratitude. As Jesus said, "a good tree bears good fruit" (Matt. 7:17). The other side (grace) is primary and is the source of faith; the standard theological definition (unmerited divine favor) is a good one.

Therefore having been justified by faith, we have peace with God through our Lord Jesus Christ (Rom. 5:1).

Being justified as a gift by His grace through the redemption which is in Christ Jesus (Rom. 3:24).

By grace you have been saved through faith; and that not of yourselves, *it is* **the gift of God; not as a result of works, that no one should boast (Eph. 2:8-9).**

In this famous passage that spells out forever more that salvation is God's free gift, Paul tells us something more. By the language used, it is clear that the "gift" of God referred to is the entire *act* of salvation. Thus, Christianity stands absolutely alone among the world's religions in this type of salvation. Here, the believer receives the ultimate gift. Like Martin Crane, we indeed want to "pay God back"; we independent Americans especially love to "pay our own way." But all such attempts to "deserve" salvation are doomed to failure. Nobody earns salvation by following the Law, or any moral code whatsoever. Salvation is a gift from God: through faith in His divine Son, we are declared righteous, holy, and *perfect*. To try to improve on such perfection is not only arrogant but logically impossible!

RESURRECTION

To understand the Christian idea of resurrection, one must understand the historical context of what competing worldviews thought about the afterlife. Remember Paul's trip to Athens in the book of Acts? Turn to the story (Acts 17:16-34) and read it.

Paul is presenting the gospel in the very heart of diverse religions and philosophies. During its heyday, Athens was the home of the famous philosophical school of Plato and Aristotle. Although Athens is somewhat past its philosophical prime when Paul visits, it is still the meeting place for ideas. For an intellectual, it remains *the* place to be.

If you look at Paul's message, he seems to be doing okay until he mentioned Jesus' being resurrected from the dead (v. 31). At that exact point, the text reads, "some began to sneer" (v. 32). Why? Why did this particular point cause an abrupt end to Paul's speech?

The answer is quite simple, really. Greek philosophy had a long history of disliking the physical realm and matter (including one's body) and ascribing primacy to the spiritual. Plato (b. 428 BC), for example, called the body "the prison house of the soul." The body held one back, not only from the next life, but, with its passions and desires, from the pursuit of wisdom in this life. Thus, in Athens, the headquarters of Plato's academy, there existed a strong "anti-body" and "anti-matter" prejudice if you will. Such prejudices were so strong, in fact, that Greek theories of creation often placed quasi-divine intermediaries as the actual creators of the world, believing that *it was unthinkable that a god or gods would dirty their hands with physical matter.*

How different this is from the Hebrew/Christian tradition of God directly creat-

> It was unthinkable that a god or gods would dirty their hands with physical matter.

ing the physical universe. Not only did God not worry about getting "contaminated" by creating the world, but he pronounced His creation "very good" (Gen. 1:31). Thus, the Hebrew soil, from which sprang Christianity, always maintained a healthy respect for the physical world in general, and the human body in particular.

> Clearly, the Bible sees the resurrection of Jesus as the pattern for all Christians.

True, in Judaism there were minority parties such as the Sadducees who denied the resurrection (read Acts 23:6-8), but such were outside the mainstream Jewish tradition. Clearly, the majority party, the Pharisees, believed in the resurrection of the body. Furthermore, despite the lack of clear Old Testament teaching on the subject, there had developed during the first century a considerable tradition that utilized elaborate theories of the resurrection of the just. The apocryphal literature written during this time period is known for its emphasis on the bodily resurrection as God's tool to thwart evil. Bad kings and rulers may slay the righteous, but God's raising them to life again only demonstrates His sovereignty and power.[2]

Thus, when it comes to the Christian concept of resurrection, it is *always* a resurrection of the body that is in view. Therefore, when Paul attempted to expound on that doctrine in Athens, he was met with scorn and ridicule. No wonder! The last thing the Greeks wanted to hear was that this new religion (Christianity) would give them a resurrected *body*! They were no doubt horrified. However, this episode teaches us a very important lesson regarding resurrection: the Christian idea always includes with it the *physical* resurrection of the body, not some ethereal spiritual resurrection. With this in mind, let's examine four key passages regarding resurrection, dealing both with Jesus' own resurrection and that of the believers. *Clearly, the Bible sees the resurrection of Jesus as the pattern for all Christians,* the "first fruits" of the believer's own resurrection (1 Cor. 15:20).

Jesus answered and said to them, "Destroy this temple, and in three days I will raise it up." The Jews therefore said, "It

took forty-six years to build this temple, and will You raise it up in three days?" But He was speaking of the temple of His body (John 2:19-21).

Memorize this verse. Know its context. It's placed first here, due to its occurring first in the life of Jesus. The Jews had asked Jesus for a miracle or "sign" to vindicate His overturning the tables of the moneychangers in the temple. Jesus replied that He would give the ultimate sign or proof of His identity as messiah: using the word "temple" as a metaphor for His body, Jesus claimed that He would rise again, three days after His death. Lest any reader misunderstand what Jesus had in mind, the Apostle John adds the editorial comment: "But He was speaking of the temple of His body" (v. 21). Thus, this verse, written by an eyewitness to Jesus' life and ministry, one who knew Him intimately, is clear: Jesus not only predicted His resurrection, but He foretold that it would be a *physical* resurrection, a raising of *His body*.

John goes on to say (v. 22) that after the resurrection took place, Jesus' disciples remembered *two* things about it: 1) that Jesus had predicted it; and, 2) that the "Scripture" had foretold such a resurrection. At this point, most Bibles refer the reader back to Psalm 16:10. Here, David speaks of both his own death and the Messiah's, and that God would not abandon David's soul to the grave (thus, his spirit would be with the Lord), and that *God would not let His "Holy One" (clearly a reference to the Messiah) undergo decay (thus speaking to a physical resurrection).*

He Himself stood in their midst. But they were startled and frightened and thought that they were seeing a spirit. And He said to them, "Why are you troubled, and why do doubts arise in your hearts? See My hands and My feet, that it is I Myself; touch Me and see, for a spirit does not have flesh and bones as you see that I have . . . they gave Him a piece of a broiled fish; and He took it and ate *it* before them (Luke 24:36-43).

> God would not let His "Holy One" undergo decay.

This is the classic description of the resurrected Jesus. You've probably read it dozens of times. What things stand out about it now? First, Jesus knew that the disciples thought He was a spirit. He gave them three specific proofs that His resurrection was indeed physical. 1) By showing them His "hands and feet," Jesus is telling the disciples to look for the telltale spike wounds of the crucifixion. Thus, it's not just any body standing in front of the apostles, it's *Jesus' own* body; 2) He let them touch Him, giving them the chance for empirical proof, and stating that His body had "flesh and bones" just like theirs (and unlike a spirit); and 3) He actually ate food in front of them. *It is clear from the text that Jesus ate the fish to quiet and calm their fears*, and to convince them that, yes, it was their Master who stood before them. Thus, the common, everyday action of eating was meant to show that however God had worked this miracle out, it was Jesus "in the flesh" that had defeated death and who talked with them now.

> It is clear from the text that Jesus ate the fish to quiet and calm their fears.

Then He said to Thomas, "Reach here your finger, and see My hands; and reach here your hand, and put it into My side; and be not unbelieving, but believing" (John 20:27).

These are the famous words and appearance which provoke Thomas's strong affirmation of Jesus' deity. Without doubt, in words impossible to misconstrue, Jesus proves that it was not only a physical resurrection, but a resurrection of the particular body of Jesus. Like the passage in Luke, Jesus had shown the marks of the crucifixion to the disciples (v. 20) and to Thomas (v. 27).

What is the purpose of a *physical* resurrection? Well, as mentioned, the Old Testament had predicted that it would be physical, but such a reply doesn't really answer the question, it just takes it further back. A physical resurrection is ultimately necessary because it is *all* of the creation that must be redeemed. Take a quick look at Romans 8:22. Here, Paul says that sin has affected the entire creation to the extent that it "groans," awaiting its redemption (vv. 21-23). Thus, cougars, pine trees, beetles, swans, spaniels, all of the creation in its physical nature, from dust and earth to bone and sinew await redemption. We ought to know this, claims Paul, for we groan inwardly with our own aches and pains, anxiously awaiting our own "redemption of our bodies" (v. 23). Thus, a physical resurrection is the result of a promised physical redemption that demonstrates once and for all the goodness of God's creation, in fact, the goodness of God.

The importance of Christ's physical resurrection points to the believer's own resurrection. For the Apostle Paul, *the resurrection is an absolutely essential part of the Christian gospel*, the "good news." In 1 Corinthians 15, Paul claims that Christ's resurrection is of "first importance" (v. 3), part of the tradition that Paul had "received" (v. 3). In fact, Jesus' resurrection is essential — not only to give ultimate meaning to this life's troubles and sorrows (v. 19) — but as a type of our own resurrection. Christ's resurrection is a "first fruits," a type of resurrection that "guarantees" the believer's *own* resurrection.[3]

Thus, the term "first fruits" refers primarily to divine, eschatological order: Christ has been raised first, then the believers at the time of Christ's coming (v. 23). Such a resurrection is part of God's plan to eventually establish His rule over all the creation (v. 25). It is clear that Paul's emphasis on the resurrection is part of his reassurances to the Corinthians — because Christ has been raised bodily, you can count on your own physical resurrection as well.

In terms of knowing of what *exactly* these resurrection bodies will consist, Paul has little patience for such speculation. He is insistent that, yes, they are *physical* bodies (else all his analogies in verses 36-49 would make no sense), but they are as different from our present bodies as the wheat stalk is from the wheat grain (vv. 37-38). Look at the following comparison.

> **The resurrection is an absolutely essential part of the Christian gospel.**

Earthly Bodies	Heavenly Bodies
Similar to a simple grain	Similar to the mature plant
Perishable	Imperishable
Sown in dishonor	Raised in glory
Sown in weakness	Raised in power
Natural body	Spiritual body

Keeping in mind Paul's absolute insistence on the Christian doctrine of the resurrection of the *body*, it is worth pointing out that Paul's thrust is not to say that "man possesses a body," but that he *is* body."[4] Will the postresurrection body be different than our present body? Most certainly! But nevertheless, it will be recognizable as *our* body, it belongs *with us*.

Any objections to the physical nature of the resurrection can be easily dealt with. Such objections center around the strange qualities of this new body. In other words, "If Jesus was raised in a physical body, why is such a body able to disappear (as to the disciples at Emmaus — Luke 24:31) or pass through matter such as doors (when He appeared to the disciples (John 20:19,26)?"

First of all, the resurrection bodies undoubtedly have properties unknown to our present human situation. With all the new qualities ascribed to these bodies (see above), shouldn't we expect that such differences in description would result in differences in capabilities? Second, such objections always sell God far short in His dealings with the "ordinary" physical realm. Such "miraculous" capabilities as exhibited by Jesus' resurrection body pose little or no problem to the God for whom "all things are possible." Remember the story of Philip and the Ethiopian Eunuch (Acts 8:26-40)? Philip's body was as ordinary as our own, but the Scripture indicates that the Holy Spirit "snatched Philip away" (v. 39). In other words, as far as the Eunuch knew, Philip disappeared! Thus, the only really important question is whether or not there is a God Who is actively involved in human affairs. If such is the case, then we should expect strange and wondrous things to happen — *He who created and formed matter would have little difficulty manipulating it!* Since He can cause our current human bodies to behave miraculously, we have all the more reason to suspect He can do so with our resurrection bodies.

THE SECOND COMING

[Jesus] was lifted up while they were looking on, and a cloud received Him out of their sight. And as they were gazing intently into the sky while He was departing, behold, two men in white clothing stood beside them; and they also said, "Men of Galilee, why do you stand looking into the sky? This Jesus, who has been taken up from you into heaven, will come in just the same way as you have watched Him go into heaven" (Acts 1:9-11).

When I was a boy I had absolute criteria for denoting what was a good birthday or Christmas gift. Hard, heavy gifts — preferably that rattled when shaken — were always good signs that this particular gift would rate high on my approval scale! Soft boxes caused groans; they inevitably held ties, handkerchiefs, or sweaters — items largely unappreciated by boys too young to shave. One Christmas I was fooled. Opening a heavy box, I was surprised to find a heavy woolen sweater. With the requisite groan, I quickly tossed it aside, only to catch part of the wool yarn on the metal eyelet of my boot. Before my mother could protest, I grabbed the offending part of yarn and gave it a hard yank. Much to my surprise (and my mother's horror), a good part of the sweater began to unravel! Needless to say, the sweater was ruined, and my parents taught me a whole new way of opening gifts.

> He who created and formed matter would have little difficulty manipulating it!

It is helpful, I think, to look at Jesus' return to earth much like that woolen sweater. *There are crucial elements to the "fabric" of how Christians view the return of Christ.* To disbelieve in one is to "pick away" at the yarn that holds the whole doctrine together. Or, to put it another way, these various elements fit together to form one coherent whole. Only at our peril do we neglect or alter one element. Each is necessary. Let's analyze the various components of the second coming of Christ: the nature, the timing, the imminence, the dangers, and the purpose.

> **There are crucial elements to the "fabric" of how Christians view the return of Christ.**

Jesus will come back bodily and physically. As the century and millennium have wound down, the world has witnessed all sorts of rumors and stories regarding the second coming of Christ. Various cults have gathered their faithful together, believing their destiny will arrive on a comet; others have pinned their hopes on various world events, certain that the latest newsworthy event in the Middle East is triggering the "last days." Historically (and sadly), when such "millennium fever" broke out, there have been Christians who sold all their belongings, quit their jobs and left their loved ones, and then hiked up mountains to be closer to heaven when Jesus came!

Keep two things in mind. One, as we enter the new millennium, such absurdities will no doubt increase. And two, such stories will prove to be utter nonsense, based upon faulty theology, philosophy, and exegesis. Nevertheless, it is indeed true that one of the most important doctrines of the Christian faith concerns the second coming of Jesus. All Christians yearn to meet their Lord face to face — it is certainly the heart-cry of the Apostle John in his final recorded prayer (read Rev. 22:20). I certainly don't intend to deal with the exact *timing* of His coming, and how it might fit in with various prophecies and current events. Besides being the point of endless squabbles and disputes, such issues are clearly peripheral to the faith. What is essential, however, is the truth that Jesus will come back to earth — physically, suddenly, triumphally, and in great power.

But why is such a truth essential? Can't one be a "Christian" without buying into the idea that Jesus is coming again? Well, again, we're faced with the technical possibility of a "yes," but with abundant red flags fluttering in the breeze. For one thing, much of the Bible's unfulfilled prophecies concern Jesus' second coming to earth. If that event is in doubt, then what of all the other promises in Scripture? Jesus Himself spoke volumes of his coming back to earth. If *those* promises are empty, on what basis can one believe *any* of His promises? Relatedly, remember how we discussed how all Christian doctrines are related? Consider Jesus' resurrection, and recall how I stressed its *physical nature*. Think of how that fits in with the second coming of Jesus. Now take a look at the following scenario.

The resurrection has taken place. Jesus has appeared in different settings to various people over a period of forty days. He then leads His disciples for the last time to Bethany, just outside Jerusalem. He blesses them, and then, the Scripture reads, "He was lifted up" from them, and a "cloud received Him" from their sight (Acts 1:9). Thus, He ascended into heaven — bodily and visibly. This is the same Jesus who appeared in the upper room, eating to prove his bodily existence; the same Jesus who challenged Thomas to touch and feel his wounds. *When the angel tells the disciples that "He'll come back in the same way He left," then we know we can expect a physical and visible return.* If there were any doubt, surely the words of Jesus should settle this issue: "For just as the lightning comes from the east, and flashes even to the west, so shall the coming of the Son of Man be" (Matt. 24:27). Let's see what else Scripture can teach us about the second coming.

> When the angel tells the disciples that "He'll come back in the same way He left," then we know we can expect a physical and visible return.

How about the timing? "But of that day and hour no one knows, not even the angels of heaven, nor the Son, but the Father alone" (Matt. 24:36).

Not only cults, but more Christians have gotten into trouble over this one issue than any other issue in eschatology (the study of last things). *The clear teaching of Scripture has for some teachers apparently been no match for the tempting issue of trying to figure out just when Jesus is coming back.* Promising careers have been ruined, and worse, peoples' faith destroyed, all because some preacher or leader thought he had correctly discerned "the signs of the times" and knew something even Jesus didn't know during His time on earth.

It is worth making the point that at every key juncture in time, so-called "prophets" predicted that Jesus was most certainly coming back that year. Such is the case when every century turns over; and such fervor reaches a fever pitch when millennia change. If you've kept up in current events, you no doubt know that 1999 has been no different.

> **The clear teaching of Scripture has apparently been no match for the temptation to set a date when Jesus is coming back.**

But Scripture is clear about the foolishness of "setting dates." Jesus tells us that neither angels nor "even the Son" are privy to the date of His return. It's amazing that some people today think they have it all figured out. About the only thing to be said for any date claimed to be *the* one is the fact that now we know when Jesus *won't* return! We would do well to heed Jesus' words: "It is not for you to know times or epochs which the Father has fixed by His own authority" (Acts 1:7).

Be ready. Another critical facet of the second coming is what theologians call the "imminent return" of Jesus. This concept is related to the timing of His coming; simply put, since we can't know the time of His return, it could be anytime! One of the synonyms of imminent, "impending," comes close to the theological meaning of the word. Because Jesus could come at any time, we ought to live like it.

When Jesus tells the story of the thief breaking into the house, His point to the disciples (and us) is plain: "You too, be ready; for the Son of Man is coming at an hour that you do not expect" (Luke 12:40). So also is His story of the master who comes home unexpectedly: "The master of that slave will come on a day when he does not expect *him*, and at an hour he does not know" (Luke 12:46).

Immediately after warning the disciples that "only the Father" knows the time and date of the Son's return, Jesus adds an admonition for watchfulness: "Be on the alert, for you do not know which day your Lord is coming. . . . For this reason you be ready too; for the Son of Man is coming at an hour when you do not think *He Will*" (Matt. 24:42,44). Paul warns the Thessalonians that "the day of the Lord will come just like a thief in the night" (1 Thess. 5:2). These warnings about Christ's return are all linked to a proper attitude of watchfulness. Remember the parable of the Ten Virgins? The wise virgins were ready when the bridegroom returned unexpectedly; the foolish ones were not. Again, Jesus ends the story with "Be alert" (Matt. 25: 1-13).

False Christs. Since there is this aura of mystery about the return of Christ, it is no wonder that there have been imposters, those announced as "the Christ" — either self-proclaimed or so designated by their misguided followers. Jesus again cuts to the heart of the issue and, in essence, takes away the guesswork. "If anyone says to you, 'Behold, here is the Christ,' or 'There He is,' do not believe *him*. For false Christs and false prophets will arise" (Matt. 24:23-24). He goes on with more warnings, and, in reality says, *"Nobody will have to tell you when the real Christ returns, you'll know!"* (Matt. 24:23-27).

Such a simple and straightforward reading of Jesus' return is far different from the picture painted by many fringe religious groups. I remember writing a paper in graduate school on Benjamin Crème, a self-described prophet of the New-Age movement. Back in 1982, he was sure that the "Maitreya," the New-Age name for Christ, was alive and well and living in London! According to Crème, the Maitreya was waiting until the timing was just right — then he would reveal himself as the Christ, the one Christians call Jesus. We could be in for a long wait!

> Immediately after warning against datesetting, Jesus adds an admonition for watchfulness.

The Purpose of His return. The final piece to the puzzle concerns the purpose for Jesus' return — *why* is He coming back? Again, let's start with the words of Jesus Himself. "And all the nations will be gathered before Him; and He will separate them from one another, as the shepherd separates the sheep from the goats" (Matt. 25:32). This is a far cry from the tiny baby in Bethlehem, meek and mild. Jesus returns now as the judge of humankind.

> Nobody will have to tell you when the real Christ returns, you'll know!

Part of such judgment is to bring into the open the secret motives behind actions. Paul says that Jesus will "bring to light the things hidden in the darkness and disclose the motives of *men's* hearts" (1 Cor. 4:5). The other part of the judgment is less personal and more cosmic. Jesus is bringing judgment upon the world, affecting every aspect of the creation. Peter says that the Day of the Lord will usher in universal destruction, with the heavens "disappearing" and the "elements destroyed" (2 Pet. 3:3-4,8-10).

It would be a mistake, however, to infer that Jesus' return is all gloom and doom. Part of His judgeship involves the reward of good deeds. Paul calls Jesus the "righteous judge" who will reward the apostle with a "crown of righteousness" at His coming (2 Tim. 4:8). Thus, believers look forward to Jesus' return as part of the consummation and ratification of their entire ministry to Him.

Thus, we have completed the fabric of the second coming of Jesus. I have tried to show that to pull one thread of the garment is to risk destroying the entire whole. If one believes, for example, that Jesus has already returned, then such a belief would fly in the face of *several* of these key components. For one thing, it makes a mockery of Jesus' words that His coming would be like lightning that flashes from one end of the sky to the other (Matt. 24:27), or that several cosmic signs would precede or accompany His return (2 Peter 3:3-4,8-10). It also completely ignores that Jesus' return will be both bodily and visible. If Jesus has returned, *Where is He? What is He doing? How exactly is He judging anything?* These questions are virtually unanswerable for anybody claiming both that they are faithful to the Bible, and that Jesus has already returned.

We have now completed the study of the foundational doctrines of the Christian faith. Let us now begin to examine the doctrines of the Jehovah's Witnesses, recalling what we've learned about the importance of the cardinal doctrines.

---

[1] See the very interesting article in Brown, *NIDNTT*, Vol. 3, p. 779.

[2] See George Nickelsburg, *Jewish Literature between the Bible and the Mishnah* (Philadelphia: Fortress Press, 1987), pp. 18, n. 5, 120-121, 254-255.

[3] Brown, *NIDNTT*, Vol. 3, p. 417.

[4] Ibid., Vol. 1, p. 235.

REFLECTING ON LESSON THREE

1. Why is the doctrine of Jesus' return so essential to a proper understanding of the Christian faith?

2. What is the Law's relation to the doctrine of salvation?

3. Name the various components of salvation. With which of these can you most easily identify?

4. How does the Greek idea of the afterlife differ from the Judeo-Christian viewpoint?

5. What are some key passages that emphasize the physical nature of Jesus' resurrection body?

6. If Jesus' resurrection body was physical, how could it disappear?

7. Discuss the different components of the return of Christ. Have you heard any stories or rumors regarding Jesus' return? How does your new knowledge affect such rumors?

RESOURCES FOR FURTHER STUDY

Christian Theology: An Introduction by Alister McGrath. A great resource that links the study of Christian doctrines with their historical roots.

FOUR

WHAT JEHOVAH'S WITNESSES BELIEVE

This chapter is designed so you the reader can easily and quickly compare the doctrines of Jehovah's Witnesses with those of historical Christianity. As we examine these beliefs, keep in mind that this book does not cover *every* Witness doctrine — it focuses on those cardinal doctrines covered in chapters 1–3. For further reading into the beliefs of JWs I suggest you utilize one or more of the additional sources listed at the back of the book.

FIRST THINGS FIRST

When analyzing any religious group, one must be careful to be as objective and as accurate as possible. Not only is it academically dishonest to let one's biases influence how the subject is reported, but objectivity is especially important when studying a subject as personal and as sacred as religion. In this same vein, accuracy is essential as well. It is crucial to "get it right" when it comes to reporting the facts of various belief systems. Toward that goal, in reporting what Jehovah's Witnesses believe, I have gone to their official sources whenever possible. Thankfully, the Watchtower Bible and Tract Society, the official publishing organ of Jehovah's Witnesses, has been quite forthright in stating its views.[1]

A Brief History

During the latter half of the 19th century, religious fervor ran rampant in America. The country was just coming to grips with the fact that it had lost over 600,000 men in the Civil War — a nearly unbelievable number, and a huge percentage of the country's population. Such facts forced at least a reexamination of the unbridled optimism that had reigned until then. Was this great conflict one of the "wars and rumors of wars" spoken of by Jesus and which would signal His return to earth? Industry and technology were drastically changing people's lives as well. Covered wagons rumbling over the Oregon Trail were becoming a thing of the past — one could now be whisked luxuriously across the country by rail in a matter of days rather than months! Strange new inventions were also making their appearance in the homes of the wealthy — lights that could seemingly burn forever and which needed no kerosene because they used something called "electricity," "talking machines" that allowed people to actually hear other people's voices over hundreds of miles. Glory! What would happen next?

> Various Adventist groups fueled messianic expectations and filled large lecture halls.

In this period of turmoil and upheaval, there were a whole host of religious groups that speculated that Jesus was coming back soon — certainly before the end of the century. Under the umbrella name of "Second Adventists," *these various groups fueled messianic expectations and filled large lecture halls*. Charles Taze Russell, the founder of what would later become the Jehovah's Witnesses, was influenced in his youth by such groups and speculations.[2] He had read with interest the prophecies of William Miller who had taught that Jesus would come back either in 1834 or 1844. Although recognizing that Miller had been wrong in his prophetic "date-settings," Russell himself was caught up in the near hysteria of the times, noting that Miller had served a useful purpose in preparing the Lord's people for His return.[3]

In point of fact, it was not really Miller's date-setting that upset Russell, but the *manner* of the return prophesied by Miller and others. The vast majority of the Second Adventists held to the orthodox doctrine that Jesus would return to earth visibly and physically. Russell, however, fastened onto a belief peddled by N.H. Barbour, a newspaper editor who claimed that Christ had returned *invisibly* in 1874.[4] Such a belief in the invisible or *spiritual* return of Christ accomplishes two things: it begins to distance Russell from historic Christianity, and it provides a ready explanation for failed prophecies.

It is important to note that the founder of Jehovah's Witnesses, like those who founded Mormonism, Christian Science, and many other non-Christian groups of the era, was a person of little formal education. With no college training, and certainly no seminary or Bible College background, *Russell was easy prey for various and strange religious beliefs.* He had already copied most of his beliefs on Jesus' return from Miller and his followers; he had only to add his own peculiar interpretations.

Since, according to Russell, Jesus had returned invisibly in 1874, and his spiritual work would occupy him exactly three and a half years, Russell gave the first (by no means would it be the last) prophecy concerning Jesus' return. He announced that Jesus' glorification and the setting up of the Kingdom of God on earth — Christ's *physical* manifestation — would take place in 1878. When that year came and went, Barbour, the original founder of the "spiritual return theory," became discouraged. But, by all accounts, Russell simply pressed the spiritual explanation into service once more. Thus began a long history of Russellite date-setting; and, when such dates came and went without Jesus' return, an equally long history of revisionist explanations by the organization of Jehovah's Witnesses.[5]

> **Russell was easy prey for various and strange religious beliefs.**

As mentioned earlier, Russell had no theological education, and he evaluated Christian beliefs by whether or not they *made sense* to him. This individualistic and peculiar use of **reason** to evaluate the truth of theological doctrine is a legacy of Charles Taze Russell bequeathed to modern-day Jehovah's Witnesses. It caused Russell to disbelieve the historical doctrine of the Trinity, the doctrine that humans had an immortal soul, and the historic Christian doctrine of hell.[6] According to Russell, such doctrines were simply not reasonable. From Russell to the second President of the Watchtower Bible and Tract Society, Joseph Rutherford, all the way to the current leadership of the Jehovah's Witnesses, the standard by which all doctrines are gauged is their comparison to reason — the reason of those at the organizational headquarters.

> The individual JW is entirely dependent for his or her "theological education" on the literature pipeline from Watchtower headquarters.

It is simply a matter of fact that the Watchtower organization employs no known biblical, theological, historical, or linguistic scholars. Thus, the standard method of developing theological doctrine within Christianity — utilizing sound biblical exegesis, and comparing it to the exegesis of those Christian scholars who have gone before, is entirely absent in the structure and teaching of the Witnesses. *The individual JW is entirely dependent for his or her "theological education" on the literature pipeline* that has its source at the New York headquarters of the Watchtower Bible and Tract Society. And, it must be said, this pipeline is entirely anonymous. As is the case with newspaper editorials, all the doctrinal statements and articles of the Watchtower organization are unsigned, with thus no way to question the author's scholarship, no way to engage the organization itself in fruitful theological discussion. All one can do is examine the doctrines of the Jehovah's Witnesses, and contrast them with their equivalents in historic, orthodox Christianity. Any discussions of doctrine are thus left to the "doorstep dialogues" with which most of us are familiar.

JESUS AND THE JEHOVAH'S WITNESSES

Jehovah's Witnesses deny the true deity of Jesus, and have assigned Him their own peculiar identity. According to Watchtower doctrine, Jesus is "God's first creation," "not equal to God in any way," and "not God's equal."[7] *Jesus is described in JW literature as "a created spirit being, just as angels were spirit beings created by God. Neither the angels nor Jesus had existed before their creation."*[8] Here, the Jehovah's Witnesses sound very much like the heretic Arius of the fourth century who said that "there was a time when the Word [Jesus] was not."

Since the JWs deny that Jesus is the Lord God, then who do they say He is? Their own translation of the New Testament, *The New World Translation*, calls Jesus "a god"; and Jesus is compared to angels in JW literature (above). Do the Witnesses then make Jesus into an angel? They claim that "Jesus has a *position* far higher than angels, imperfect men, or Satan," and that Jesus is a "Mighty God."[9] But the truth is that Jehovah's Witnesses think that Jesus is none other than Michael the Archangel! Look at the following quotations from JW literature:

"**The foremost angel, both in power and authority, is the archangel, Jesus Christ, also called Michael.**"[10]

"**War broke out in heaven. Michael (Jesus Christ in heavenly power) and his angels battled with the dragon.**"[11]

"**Reasonably, then, the archangel Michael is Jesus Christ.**"[12]

If they describe Jesus as being "higher" than angels, it is only positionally higher (see above), much like an ordinary citizen being appointed to a high government post. He may be *positionally* higher than his former cronies, but entirely equal in nature, being, and substance to the citizens from whence he came. Thus, for JWs Jesus Christ is an elevated angel.

This identification of Jesus as Michael puts the Jehovah's Witnesses in a diffi-

> Jesus is described in JW literature as "a created spirit being, just as angels were spirit beings created by God."

cult position. Anytime they claim to "believe in Jesus," or that "Jesus died as a ransom," or that "Jesus is my savior," they are putting their faith in an angel to save them. It also raises troubling issues for their version of Christ. In Jude 9, Michael the Archangel is disputing with Satan over the body of Moses. The text is clear that Michael "did not dare" pronounce judgment upon Satan in his own name, but instead, used the Lord's name to rebuke Satan. If Jesus is Michael, then why did Jesus in his earthly ministry rebuke Satan Himself (Matt. 4:10), and this during the time when the Scriptures say that Jesus was a little *lower* than the angels (Heb. 2:9)? Why did he exercise so much of *his own* authority while on earth if he was *lower* than his previous angelic state — when he could *not* do so?

> The angels are not heirs, sons, objects of worship, kings, or Lord — all titles or attributes given to the Son Himself, Jesus Christ.

Even more pointedly, the entire first chapter of the book of Hebrews *sharply* contrasts Jesus with the angels. It is **the Son** who is described as a) heir of all things (v. 2); b) the creator of the world (v. 2); c) the radiance of the Father's glory (v. 3); d) the exact representation of the Father's nature (v. 3); e) the upholder of all things (v. 3); f) **much better than the angels (v. 4)**; g) inheriting a more excellent name than the angels (v. 4); h) the object of worship **by angels (v. 6)**; i) God Himself (v. 8) (emphasis mine). The angels, on the other hand, are ministers or ministering spirits, sent out to render service to believers (vv. 7,14); *they are certainly not heirs, sons, objects of worship, kings, or Lord — all titles or attributes given to the Son Himself, Jesus Christ.* In fact, it is obvious that the entire point of the first chapter is to refute the notion that Jesus is an angel. According to the writer of Hebrews (1:8), who is quoting Psalm 45:6, Jesus is indeed Jehovah God!

This worship of the Son of course raises yet another problem for JWs. Why in the world would the writer of Hebrews claim that "*all* the angels of God" are to worship the Son (1:6)? Angels know full well that they must only worship

God (Rev. 19:10; 22:9). The message is clear: the Son deserves worship because He *is* God!

On a more earthly note, Jesus is worshiped not only by angels but by men. The Scriptures are absolutely clear that after Jesus walked on the sea, the disciples "worshiped" Him (Matt. 14:33). Since Jesus knew full well that only the Lord God was to be worshiped (Matt. 4:10), His acceptance of His disciples' worship points unequivocally to only one conclusion: that *Jesus knew He was worthy of worship, that is, He was God!* The Jehovah's Witnesses' almost ludicrous attempts to explain away the disciples' and angels' worship of Jesus are tortuous at best, disingenuous at worst. Admitting that the word used to designate the worship toward Jesus is the same as that used by Abraham toward Jehovah (Gen. 22:5), they bring in instances where the word is used "to show respect" or "to honor those with whom one does business."[13] Their unspoken assertion? That God was only commanding the angels to show His Son some respect! Likewise, it stretches human credulity to the breaking point to believe that the disciples had just witnessed Jesus feeding 5,000 people with a few leftovers, then walking on the raging waters of a storm on the Sea of Galilee, then halting the violent tempest in its tracks, then bowed down to Jesus in order to show Him "a little respect!" No, the context is unwavering: they worshiped Him as they would worship the Lord God.

This foundational error in Christology by Jehovah's Witnesses forces them into other strange interpretations of Scripture. Not only do they believe that Jesus was formerly called Michael before his birth in Bethlehem, they hold that Jesus *is now* Michael once again! "So the evidence indicates that the Son of God was known as Michael before he came to earth and is known also by that name since his return to heaven."[14] But, if Jesus reassumed his identity as Michael after His ascension to heaven in Acts (1:9), then why does he identify himself as *Jesus* to Stephen the Martyr (Acts 7:55), the persecutor Saul (soon to become Paul — Acts 9:5), and to the Apostle John (Rev. 22:16)? In

> **Jesus knew He was worthy of worship, that is, He was God!**

fact, in the last two instances, Jesus *personally* identifies Himself *as Jesus*. If He were Michael, such contexts would have been the perfect opportunity to declare himself such — but he did not.

> **It is Jesus — not Michael — in whose name alone is salvation.**

In fact, one is forced to wonder why, since Jehovah's Witnesses believe that Jesus is really Michael now, they make mention of Jesus at all! Is Jesus now known by two different names? What can they make of the insistence of Peter that *it is Jesus — not Michael — in whose name alone is salvation:* "for there is no other name under heaven that has been given among men, by which we must be saved" (Acts 4:12)? Clearly, Peter thought that it is Jesus, not Michael, who *now* acts to save those who are His. Relatedly, Peter acknowledges the full deity of Jesus when he links the name of "the Lord" in his first sermon (Acts 2:21), referring back to Jehovah in Joel 2:32, to the name of Jesus Christ in his defense to the priests (Acts 4:12). For Peter, Jesus is the God of the prophet Joel, the one who was to send His spirit upon all humanity.

When the JWs identify Jesus as Michael, an angelic "glorified spirit,"[15] they also are at odds with the Apostle Paul's identification of the current *nature* of Jesus. In his letter to Timothy, Paul states clearly that "there is one God, and one mediator also between God and men, *the* **man** Christ Jesus (1 Tim. 2:5, emphasis mine). Note the present tense of the verb "is." Writing long after the ascension, Paul claims that Jesus is not some "glorified spirit being," certainly not "the archangel Michael," but the *man* Christ Jesus. When the JWs deny this manhood of Jesus, they also betray their misunderstanding of the resurrection, but for now let's stay close to the issue of the person of Christ and the Trinity. After reading how the Jehovah's Witnesses view the Trinity, it will be clear to the reader that they have fundamentally misunderstood this important Christian doctrine.

JEHOVAH'S WITNESSES AND THE TRINITY

Since the JWs deny the deity of Christ, it follows they must deny the doctrine of the Trinity. It seems as if *two fundamental misunderstandings of the Christian doctrine of the Trinity lie at the root of their denials.* One might be put this way: "Separation of persons means inequality of persons." Examine the following quotes from the Witnesses' "Is God Always Superior to Jesus?"[16]

1) "Time and again, Jesus showed that he was a creature separate from God and that he, Jesus, had a God above him, a God whom he worshiped, a God whom he called 'Father'. . . . Since Jesus *had* a God, his Father, he could not at the same time *be* that God" (p. 1).

2) "The apostle Paul had no reservations about speaking of Jesus and God as distinctly separate . . . . Here [John 8:17,18] Jesus shows that he and the Father, that is, Almighty God, must be two distinct entities" (p. 1).

3) "Paul also said that Christ entered 'heaven itself, so that he could appear in the actual presence of God on our behalf' (Heb. 9:24). If you appear in someone else's presence, how can you be that person? You cannot. You must be different and separate" (p. 5).

There are many other instances in which the Jehovah's Witnesses make this same flawed argument: since Jesus and the Father are separate persons, then Jesus cannot be God. Such an argument, however, is really addressing a straw man, and assumes that Christians believe that the Father and the Son are the same *person*. As any astute Christian will know, the Jehovah's Witnesses are not refuting the Christian idea of the Trinity at all, but rather a distortion of it. Christians have always *agreed* that Jesus and the Father are separate persons or individuals. In fact, it was an early *heresy* that held the beliefs that are now addressed by the JWs' arguments. This heresy, "Modalism," held that "Father," "Son," and "Holy

> Two fundamental misunderstandings of the Christian doctrine of the Trinity lie at the root of the denial of Jesus' deity.

Spirit" are merely names for the different "modes" of God. According to this view, God assumes the title and personage of Jehovah in the Old Testament, then changes roles in the New Testament to be known as the Son, then changes roles a third time to be known now as the Spirit. To refute this error, Christian theologians in fact used many of the exact same arguments used today by JWs against the so-called Trinity! For example, how could Jesus be the same person as the Father since He is so often depicted as separate from Him, and relating to Him (as in prayer)?

> Christians actually agree with the JWs' argument that the Son is not the Father, and that they are separate persons.

Thus, *Christians actually agree with the JWs' argument that the Son is not the Father, and that they are separate persons.* But does such separateness necessarily imply inequality of being? Of course not! You are a separate person from your relatives and friends, but you share a common human nature. This brings us to the second foundational error that JWs make concerning the Trinity: denying the true human nature of Jesus.

Christian doctrine holds that Jesus assumed a sinless human nature to complement His divine nature at the incarnation. One should not think, however, of Jesus being half God and half man, but 100 percent of both. Part of Christian theology holds that Jesus often spoke from that human nature, and as man, he could speak with the frailties and limitations (not sinfulness) of humanity. Thus, when Jesus speaks of ascending to "My Father and your Father, to My God and your God" (John 20:17), He speaks in His humanity. As the man Jesus, He worshiped the same God as Christians revere today.

Nowhere is such humanity more evident as when Jesus speaks as the "suffering servant" foretold long before by the prophet Isaiah (Chapter 53). When such instances are in view, Jesus usually takes for Himself the title "Son," emphasizing His filial role as man's submissive representative before the Father. Thus, when Jesus says, "the Son can do nothing of Himself, unless *it is* something He sees the Father doing" (John 5:19), that speaks to Jesus' incarnation as the one who lives to do the will of the Father (in contrast with Adam's failure in the Garden of Eden). Such submission is claimed by the JWs to be clear proof that Jesus is not God — for how could a submissive person claim equality with the one above him? (pp. 2, 3, 5). However, one can indeed submit to his equal. A person could voluntarily submit to the government leadership, for example, but still quite rightly claim equality of nature to them.

Since submission is a voluntary act of the will, one could even submit to those whose stature is *inferior*. Such is exactly the case when the youth Jesus voluntarily submits to his parents after their trip to Jerusalem (Luke 2:51). JWs admit that Jesus is far higher and more exalted than "ordinary people." But how can that be so if He submitted to them in the persons of His parents? Since the Witnesses equate submission with inferiority, they can have no logical reply to that.

Such is precisely the context of Jesus' ignorance of the timing of His return, that event that "only the Father" knew (Mark 13:32). Jehovah's Witnesses see in this ignorance clear proof that Jesus could not be God because, by definition, God knows everything.[17] Such arguments may sound convincing until one looks at the incarnation itself, the act of God becoming a man. During His earthly ministry as the suffering servant, Jesus took on the body of an ordinary man, a carpenter's son. Look at the apostle's words in Philippians 2:7, where *Paul talks about Jesus' "emptying Himself" by "taking the form of a bond-servant."* The **form of a servant** is in sharp contrast to the preincarnate Jesus who "existed in the **form of God**" (v. 6); thus, Paul's point is not to deny the deity of Jesus (for he elsewhere clearly affirms it — Col. 2:9), but to urge his readers to live lives of humility, because their Lord exchanged the totality of divine glory for the lowly form of a man. This "emptying" then doesn't reflect an emptying of the divine nature, but an acceptance of any limitations imposed by the human

> Paul talks about Jesus' "emptying Himself" by "taking the form of a bond-servant."

body.[18] Thus, Jesus was hungry, became tired, learned obedience (Heb. 5:8), grew in wisdom (Luke 2), and was fully part of the essential human experience (Since Adam was once sinless, we then know that Jesus' essential humanity need not be sinful). This human experience in effect limited His divine omniscience in this plane — speaking as the Son, Jesus indeed did not know the time of His return. Nonetheless, the witness of Scripture to the deity of Christ is strong and pervasive: God indeed had dwelt among us for a time (Isa. 7:14).

> The view of the Holy Spirit as a "force" simply doesn't do justice to Scripture's witness to Him, either in direct affirmation, or in inference.

Not surprisingly, JWs redefine and reinterpret the Holy Spirit. For Christians, the Holy Spirit is the third person of the Trinity, the personal God active today amongst His people. But for Jehovah's Witnesses, the Holy Spirit is "a controlled force that Jehovah God uses . . . . To a certain extent, it can be likened to electricity."[19] *This view simply doesn't do justice to Scripture's witness to the Holy Spirit, either in direct affirmation, or in inference.*

The book of Acts bears witness to a young, vibrant, and growing church. Through the ministry of the apostles, the lame and crippled were healed; the dead were even raised; and the gospel was spread with power and fervor. During this time, many people were giving everything they had to the church, and there was a spirit of community and sharing that is all too rare today (see Acts 4:32-36). People were even selling their property, then giving the proceeds to the apostles (4:36). Apparently wanting to bask in the gratitude of the apostles and believers, a man named Ananias and his wife Sapphira sold their tract of land, then took part of the proceeds and laid it "at the apostles' feet," but represented the money as the total sale price. Peter is quick to spot the lie, and God's judgment comes quickly upon the hypocrites: "Ananias, why has Satan filled your heart to lie to the Holy Spirit, and to keep back *some* of the price of the land? . . . Why is it that you have conceived this deed in your heart?

You have not lied to men but to God" (Acts 5:3,4). Thus, Peter equates lying to the Holy Spirit as lying to God.

This reverence to the Holy Spirit has its roots in the Gospels, in the words of Jesus. In fact, Jesus goes so far as to say that although all insults uttered against Him are forgivable, when one "blasphemes against the Holy Spirit, it shall not be forgiven him" (Luke 12:10). Blasphemy, as we have seen, is defined as an insult or an affront to God; relatedly, it is of course impossible to blaspheme against a "force." Thus, Jesus' reference points the way to the personality and deity of the Holy Spirit. Such is the thrust of Paul's warning not to grieve the Holy Spirit (Eph. 4:30) — only a person can be grieved, not an impersonal force like electricity!

The divine personhood of the Holy Spirit is reaffirmed consistently in the Scriptures. The Holy Spirit: **speaks and commands** (Acts 13:2), **testifies** (Acts 20:23), **forbids** (Acts 16:6), and **teaches** (1 Cor. 2:13). Such are the actions of a person, not a force. In the upper room, Jesus comforts the disciples by telling them that another Helper is coming, One who will abide with them (John 14:17), who will "teach them all things" (v. 26), who will "bear witness" (15:26), who will "convict the world concerning sin" (16:8), and who will "guide them into all the truth" (v. 13). These personal tasks and duties are backed up by the Greek grammar, the masculine pronoun being used in every reference to the Helper (John 14–16). The Holy Spirit is no impersonal "force" or "energy," but the very active third person of the Godhead.

THE CONCEPT OF SALVATION

Remember how we discussed how changing or altering one doctrine inevitably affects other doctrines? Such is the case with Jehovah's Witnesses and their denial of Christ's deity. *Since they believe that Jesus was not God while on earth, but merely a man, what Jesus accomplished on the cross reflects that belief.* "Jesus, no more and no less than

> The divine personhood of the Holy Spirit is reaffirmed consistently in the Scriptures.

a perfect human, became a ransom that compensated exactly for what Adam lost — the right to perfect human life on earth."[20] This idea of the "ransom" is very important in JW theology and spotlights yet another doctrine in which they differ from biblical Christianity. Witnesses emphasize that since Adam was only a man, then divine justice requires the death of a man, "no more, no less."

> Since JWs believe that Jesus was not God while on earth, but merely a man, what Jesus accomplished on the cross reflects that belief.

The problem is that such judicial logic must limit — by the JWs' own reasoning — the scope and power of the ransom sacrifice of Jesus. Christian theology holds that Jesus, as the God-Man, could give His life as an atonement for not only Adam's sin, but for all the sins of all the believers — past, present, and future. Only if Jesus were God could His atoning death have such power to, in the immortal words of John the Baptist, "take away the sin of the world" (John 1:29). Thus, there is a theological imperative to Jesus being the God-Man: He had to be truly human to represent humanity in its totality, and He had to be God so His death could cover all the sins of all the believers.

But by limiting Jesus to being just a mere man, the Jehovah's Witnesses must therefore limit the scope and coverage of His ransom sacrifice. Ironically, the Witnesses seem to want their version of Jesus' ransom to be broad in scope; they admit that "All who put their faith in Jesus can have their sins forgiven and receive everlasting life."[21] But on what basis? There is simply no ground for the individual Jehovah's Witness to identify with the Christian doctrine of salvation. Their Christ is not God but an angel; and when the angel Jesus comes to earth he does so, not even as an angel, but as a man only. Such a death simply cannot do the redemptive work that JWs want it to do.

In conversations with many Jehovah's Witnesses, in moments of candor, it appears that many if not most of them have an emptiness that feeds a desire to "help" God with their salvation. Perhaps this is part of the JWs' drive that has them attending meetings five times weekly, and "volunteering" at least 10 hours per month going door to door. In any event, the JWs' concept of salvation never relies on, as the Reformation put it, the concept of "grace alone." While admitting that they cannot "earn" salvation, official Watchtower teaching is always quick to add different "requirements" for salvation besides the Christian idea of personal trust in Christ.[22] The knowledgeable Christian will always recognize and reject any form of "works" as a prerequisite for salvation, knowing that God alone has done the work (Eph. 2:8,9).

THE RESURRECTION OF JESUS CHRIST

Jehovah's Witnesses believe that Jesus was resurrected as "a spirit creature," not raised in bodily form, as is the case in Christian theology. As we have already seen, JWs believe that Jesus in heaven has once again assumed the name "Michael" as the chief of all the angels, and the "glorified spirit Son of God."[23]

They admit that Jesus appeared in "physical form" after the resurrection, but claim that Jesus "materialized" those bodies as other angels had done in the past.[24] For the Jehovah's Witnesses, the stumbling block to believing in the physical resurrection of Jesus appears to be threefold: 1) He apparently was able to go through doors (and no physical body can do that); 2) Paul says "the last Adam [Jesus Christ] became a life-giving spirit" (1 Cor. 15:45); and 3) A spiritual resurrection allows them to maintain that Jesus has already returned to earth — spiritually. Let's take each in turn.

> Jehovah's Witnesses believe that Jesus was resurrected as "a spirit creature," not raised in bodily form.

Christians have always admitted that the resurrection body of Jesus had special properties. The Apostle Paul takes great pains to distinguish between the "natural" and "spiritual" bodies in 1 Corinthians 15 — but they are both *bodies.* That the resurrection body has miraculous properties should pose no obstacle to belief. In His earthly ministry, in His *natural* body, Jesus defied reason and the so-called "laws" of nature by walking on water; should we hesitate because a glorified body passed through doors or rose up into heaven? Certainly, the testimony of the Scriptures is both clear and abundant on the subject of Jesus' physical resurrection. First of all, and as already discussed, Jesus Himself predicted His physical resurrection (John 2:19-21). As a pious Jew, He knew full well that God's plan included resurrection (Matt. 22:29-32). Second, the tomb was empty. *A spiritual resurrection would have no need of a stone rolled away or a missing body* — the spirit of Jesus could have simply appeared to His disciples whenever necessary. Third, He appeared to the disciples, even *telling* them He was not a spirit (Luke 24:39)! That alone should be sufficient to disprove any notion of a "spiritual resurrection." Fourth, He deliberately asked for food and ate in front of the disciples, to prove that He was no spirit (Luke 24:41-43). Finally, in His appearance to Thomas, He challenges the disciple to touch and verify the actual wounds of crucifixion (John 20:27-29). These clear passages demolish the pretensions of the Jehovah's Witnesses that He rose as a spirit. Now moving to Paul's famous passage in Corinthians (15:35-49), we are not amazed that this resurrection body has miraculous properties: the heavenly (resurrection) body is indeed sharply contrasted with the earthly (preresurrection) body — but they are both *bodies*.

> A spiritual resurrection would have no need of a stone rolled away or a missing body.

So, what about Jesus (as the last Adam) becoming "a life-giving spirit" (15:45)? Does this contradict the testimony of the Gospels? Of course not. Paul knew the apostles well, and had traveled extensively with Luke. He was fully aware that they had all testified that Jesus was raised physically. Paul is taking great pains in his letter to the Corinthians to spell out the importance of the resurrection, and to inform his audience that their resurrection bodies will not be some simple "remakes" subject to the same illnesses and infirmities of which they were all too aware. No, their resurrection bodies would be "imperishable," "raised in glory and power," and "spiritual" (15:42-44). They would exemplify the power of the same Holy Spirit that raised Jesus from the dead. As Richard Gaffin, Professor of Systematic Theology at Westminster Theological Seminary, says, "The resurrection body of 1 Cor. 15:44 is 'spiritual' not in the sense of being adapted to the human *pneuma* or because of its (immaterial) composition/substance, to mention persisting misconceptions, but because it embodies the fullest outworking, the ultimate outcome, of the work of the Holy Spirit in the believer . . . ."[25]

The last stumbling block for Jehovah's Witnesses' believing that Jesus' resurrection was physical concerns their peculiar beliefs on the return of Christ. Since, as we have discussed earlier, JWs believe that Jesus has *already* returned spiritually, then it makes sense that, to them, Jesus could *not* have been raised physically — else how could he return *spiritually* with a *physical* body? The Jehovah's Witnesses are first and foremost an apocalyptic group that thrives on the belief that Armageddon is always just around the corner, signifying God's great displeasure with the "earthly rulers," and the setting up of God's theocratic rule on earth. Literally dozens of times, the organization has announced that this great battle was here, and that the end of the world had come.[26] And, just as regularly, the dates came and went, with no "final battle" — leaving the Watchtower organization busy revising history and their prophecies.

This vast organization simply has far too much invested to allow for a Jesus with a physical, resurrected body. For them to admit such a thing would entail no less than a demolition of one of the pillars on which the organization was structured. A Jesus who resurrected physically implies a Jesus who returns both physically *and* visibly. Such a return would be at odds with the very first principles laid down by Charles Russell — and would imply that the Jehovah's Witnesses are built upon a foundation that is *fundamentally* and *fatally* flawed. And, in the past century or so since Russell insisted upon the spiritual return, the organization has invested huge amounts of time, paper, and energy promulgating and defending just such a return. *For them to admit that this doctrine is wrong is tantamount to organizational suicide.* Thus, the Jehovah's Witnesses insist upon the spiritual resurrection of Christ — despite the clear biblical evidence to the contrary. This apocalyptic vision also gives them a unique stance in relation to society — and it is this issue we will now discuss.

> **For them to admit that this doctrine is wrong is tantamount to organizational suicide.**

---

[1] You can find most all of their official doctrines at their web site: http://watchtower.org. I have been told by official Jehovah's Witness sources that this web site represents official Watchtower doctrine. Thus, I have used documents printed from this site to compare with Christian belief.

[2] See Alan Rogerson, *Millions Now Living Will Never Die* (London: Constable, 1969), pp. 6ff.

[3] Ibid., pp. 7ff.

[4] Ibid., p. 7.

[5] Rogerson's use of primary sources proves Russell was mistaken over and over again on this count (see pp. 9-31, 77ff., 191-192).

[6] Ibid., p. 10.

[7] See "Who Is Jesus Christ?" p. 1; "Is God Always Superior to Jesus?" pp. 1, 2; found at http://watchtower.org/library.

[8] See "What Does the Bible Say about Jesus?" p. 3, found at http://watchtower.org/library.

[9] See "What about Trinity 'Proof Texts'?" p. 10. The emphasis is mine. Found at http://watchtower.org/library.

[10] See "The Truth about Angels," p. 4, found at http://www.watchtower.org/library.

[11] "How We Know We Are in 'The Last Days,'" p. 3, found at http://www.watchtower.org/library.

[12] Ron Rhodes, *Reasoning from the Scriptures* (Brooklyn: Watchtower Bible and Tract Society, 1985), p. 218.

[13] Ibid., p. 215.

[14] Ibid., p. 218.

[15] Ibid.

[16] http://watchtower.org/library/ti/god_always_superior.htm.

[17] "Is God Always Superior to Jesus?" p. 4.

[18] See Brown, *NIDNTT*, Vol. 1, pp. 548f.

[19] "The Holy Spirit — God's Active Force," p. 1, found at http://watchtower.org/library/ti/active_force.htm.

[20] "How Much Was the Ransom?" p. 5 in "What Does the Bible Say about God and Jesus?" found at http://watchtower.org/library/ti/god_and_jesus.htm.

[21] "Who Is Jesus Christ?" p. 2.

[22] See Rhodes, *Reasoning from the Scriptures*, p. 216.

[23] Ibid., p. 218. As to being raised as a spirit, see "Who Is Jesus Christ?" p. 2.

[24] "Who Is Jesus Christ?" p. 2.

[25] See "'Life-Giving Spirit': Probing the Center of Paul's Pneumatology," *Journal of the Evangelical Theological Society* 41, No. 4, p. 577.

[26] See *Studies in the Scriptures* (Brooklyn: Watchtower and Tract Society, 1909), pp. 77, 99, 101.

REFLECTING ON LESSON FOUR

1. Who is Jesus, according to the Jehovah's Witnesses?

2. What difficulties result because of the identity of the Jehovah's Witnesses' Christ?

3. What do the JWs think about the doctrine of the Trinity? What foundational error do they make to arrive at their conclusion? How would *you* state the Trinity?

4. If Jesus is God, how could He not know the time of His return?

5. How do JWs define the resurrection of Jesus? Could you reply to their beliefs? Try a little role-playing, with one person being a Witness, the other, a Christian. Can "the Christian" defend the doctrine of the resurrection of Jesus?

RESOURCES FOR FURTHER STUDY

Early Christian Doctrines by Kelly (again). Look up the issues and debates over the person of Jesus. This is a great resource for seeing how history unfolded during those times.

On the Incarnation by Athanasius. This was a classic work when written; it remains so now.

Cur Deus Homo by Anselm. The theological necessity of Jesus' being both God and man, and why it's so vital to maintain both natures.

The Orwellian World of Jehovah's Witnesses by Heather and Gary Botting. A great description of the highly organized and controlled world of JWs, it describes the source of their drive to "witness" to other people.

FIVE

JEHOVAH'S WITNESSES AND SOCIETY

This chapter deals in broad strokes, covering topics as various as how JWs view themselves and society, to their own translation of the Bible, to tips in befriending and witnessing to Witnesses. This chapter is designed first and foremost to promote understanding, without which it is nearly impossible to make any lasting connections with members of the group.

JEHOVAH'S WITNESSES AND SOCIETY — WANTING NO PART OF IT

One of the most profound books I read in seminary was H. Richard Niebuhr's *Christ and Culture*.[1] In this book, Niebuhr sets out to identify the main religious "types" in Christianity, classifying them in relation to their own perceived stance toward culture. Keeping in mind that Niebuhr was identifying *Christian* groups, and that the JWs lie outside the realm of orthodox Christianity, it is nonetheless highly instructive to use his categories in a sociological analysis of JWs.

Without a doubt, Jehovah's Witnesses are a prime example of the group Niebuhr labels "Christ against Culture."[2] This group is known for their separation from (and rejection of) what they perceive as "the world" around them. Here, society has no claims at all over the believer — his or her allegiance is solely to God and the group. When such a group examines themselves in relation to society, they love to quote Scriptures such as "Do not love the world, nor the things in the world" (1 John 2:15). *The world is viewed as totally under the power and sway of the devil*, with the members' rejecting the surrounding culture, pledging allegiance to the leaders of the group or denomination, and getting as many people as possible into the group itself. Members of this type see Scriptures that portray Satan as "the ruler of this world" as primary and overriding any apostolic commands for Christians to be good citizens (as in Romans 13).[3]

> **The world is viewed as totally under the power and sway of the devil.**

Jehovah's Witnesses are a perfect fit in this type. In "Who Really Rules the World?" the JWs claim that "nowhere does the Bible say that either Jesus Christ or his father are the real rulers of this world . . . . Satan the Devil really is the unseen ruler of the world!"[4] This claim of course is a complete fabrication, ignoring passages that reveal that God Himself owns and rules the earth (Exod. 9:29; 19:5; Ps. 24:1; 1 Cor. 10:26), but it places the Witnesses squarely in Niebuhr's type, "Christ against Culture." It also marks the JWs as taking a basically adversarial stance toward their surrounding culture and society. Such a stance plays itself out in their refusal to celebrate birthdays or Christmas, their total rejection of modern governments as "satanic," and their refusal to pledge allegiance to the flag. Indeed, JWs have a track record of persecution by various governments, all for the same basic reason — their utter rejection of any allegiance to any earthly rulers. This cultural rejection is only half the story, however.

Groups that reject society to this extent are almost always strict in their requirements for membership and belonging, and are maintained with the tightest of security and control. Such is the case with Jehovah's Witnesses. In a fascinating and compelling work, ex-Witnesses Heather and Gary Botting compare the JW organization to both the closed societies of Marxist states, and to the frightening world of George Orwell's *1984*.⁵ In a point by point comparison, *the authors present a convincing case that JWs live in an Orwellian society* where "history is re-written," doctrines are changed, members are commanded to use a "special vocabulary," the "correct" thoughts are encouraged, and life is regulated to the point that even facial expressions are monitored!⁶

In an organization so tightly regulated, punishments for breaking the tight constraints are severe. "Disfellowshipping," the practice of utterly rejecting a JW who "strays" from the truth, is the severing of all ties with JW family and friends. Such ostracizing only lends more credence to the already somber and rigid constraints of the organization. Thus, it can be of no surprise to see a striking overall parallel to the pessimistic world of Winston Smith in *1984*. Like Orwell's novel, the JWs paint the future in grim shades of gray with Armageddon lurking just around the corner of nearly every political crisis. For the JW, like the portrait painted in *1984*, the world isn't *supposed* to get better. At times, one almost senses the individual Jehovah's Witness is *disappointed* when global conflict doesn't break out!

This adversarial stance with society also has important ramifications for how the Witnesses view individual Christians. A couple of years ago in a class I was teaching on the cults one student raised her hand and asked, "Why is it, when I talk with Jehovah's Witnesses, I always seem to get so angry and frustrated?" Good question. Part of the reason is that Jehovah's Witnesses have been drilled to believe that Christians are not only enemies of the "true gospel," but they are out "to get" the individual JW as well. As mentioned, JWs have often been persecuted for their religious beliefs, with thousands dying in Hitler's concentration camps and Stalin's purges. This historical fact combined with their indoctrination against Christianity results in many JWs developing what might be labeled a "persecution complex."⁷ As I answered my student, at times they almost *expect* to get doors slammed in their faces or threats

> The Bottings present a convincing case that JWs live in an Orwellian society.

uttered on porches. After all, as they've been told, "That's what Christians are like!" To open up effective channels of communication with Jehovah's Witnesses, Christians have to use not only their heads but their *hearts*.

> **There are higher issues at stake than simply winning debate points.**

TRY BEING FRIENDS

Think for a moment on the following question: "Why is it important to talk with Jehovah's Witnesses?" If you answered, "To win arguments," you need to reflect on the message of 1 Peter 3:15: "Sanctify Christ as Lord in your hearts, always being ready to make a defense to everyone who asks you to give an account for the hope that is in you, yet with gentleness and reverence." *There are higher issues at stake than simply winning debate points.* When such stakes are examined in the context of Peter's admonition that we be "gentle" in our defense of the faith, a whole new way of dialogue opens up.

Keep in mind that I'm not advocating befriending Witnesses as some sort of sterile "strategy" designed simply as the "means to an end." What I'm suggesting is for the Christian to take a genuine interest in the Witness as a person. Next time JWs ring the doorbell, take a moment to get to know them. What are their names? Do they live nearby? Where do they work? Why did they join the Jehovah's Witness organization? Why not offer them something to drink? If you can't take the time to talk at that particular moment, make an appointment to talk at a mutually convenient time. If you don't feel ready to talk to them yourself, why not take a more knowledgeable friend along? In any case, try being friendly! Not only is such behavior consistent with Christianity, but it helps to diffuse any ideas that the JWs might have that you are going to personally attack them. And, no matter how tempting it might be, launching a loud and personal attack against Witnesses only serves to confirm what they've been taught.

TERM SWITCHING

When my student complained that frustration seemed to be the primary result of her encounters with Jehovah's Witnesses, the concept of a "persecution complex" played only part of the problem. As is the case with all non-Christian groups, JWs are experts in what Walter Martin called "term switching." That is, when talking with the average JW, I have found them adroit at using the exact same terminology as Christians but giving it an entirely different and novel meaning. *Such switching can lead to endless frustration and blocked communication.* For example, suppose a Christian and a Witness engage in a conversation — it might sound something like this:

Sally Christian: "I believe in Jesus Christ for my salvation, but you don't believe that."

Wally Witness: "Yes, I do; otherwise I would not be at your door today."[8]

Sally Christian: "Oh . . . Uh, I didn't know that. Well, I'm happy with my church."

At this point the dialogue is already in trouble. Sally is becoming frustrated because she knows — somewhere in the back of her mind — that JWs don't believe the same as Christians, but she just can't put her finger on it. And, since the Witness claimed to "believe in Jesus," she is backpedaling, and has brought up the concept "church" into the conversation — as much for self-defense as for anything else. Of course, the word "church" raises all sorts of red flags for the Witness — *they told me that Christians always bring their church into it. I'll bet she's just like all the other Christians that the clergy have brainwashed.*

What Sally needs to do is to insist that the JW define his terms. What did he *mean* by Jesus Christ? If Sally would have known that Wally's Jesus was Michael the Archangel, the conversation would have proceeded along

> **Term switching can lead to endless frustration and blocked communication.**

much different grounds, with a legitimate issue at stake.

Sally Christian: "I believe in Jesus Christ for my salvation, whom I believe to be God."

Wally Witness: "I also believe in Jesus Christ, and I also think he is a god."

Sally Christian: "But isn't it true that you think Jesus is Michael the Archangel, and a being that is inferior to Jehovah?"

Wally Witness: "Well, yes, that's true."

> Majoring on minors is never a good strategy and is a horrible tactic to employ when talking to JWs.

Now, the discussion can focus on genuine points of disagreement, with the issues out in the open. Sally can now proceed to share her faith, and the reasons why it's so important to believe in the deity of Christ.

Another point to keep in mind is the importance of sticking to the main issues, such as those discussed in the first three chapters of this text. *"Majoring on minors" is never a good strategy and is a horrible tactic to employ when talking to JWs.* Like any other religious group, JWs have a whole host of issues with which you may disagree. But, like shooting at a target, keep your points in the center, around the "bull's-eye." Yes, it is important to *know* about such tangential beliefs — it helps you learn more about the person. But Christians know that one's eternal destiny doesn't hinge, for example, on whether or not one believes blood transfusions are legitimate for the church.

WHAT ABOUT THAT BIBLE OF THEIRS?

Most cult experts claim that one of the identifying marks of a non-Christian cult or group is its insistence that the Bible is not enough, that extrabiblical materials and books are necessary to get "the whole" truth. Mormons have their *Book of Mormon*; Christian Scientists have the writings of Mary Baker Eddy; Jehovah's Witnesses are no different, elevating their own writings to canonical status.

Jehovah's Witnesses, in fact, *fear* the practice of reading the Bible by itself, without the organization's materials to "guide" the reader. In a revealing admission, the Watchtower organization claims that any Witness who reads only the Bible goes "into darkness" (believes in traditional Christianity) without the organization's *Scripture Studies*.[9] In case there remains any doubt what is meant, listen to the warnings of a recent *Watchtower*:

> They [the doubters] say that it is sufficient to read the Bible exclusively, either alone or in small groups at home. But, strangely, through such "Bible reading," they have reverted right back to the apostate doctrines that commentaries by Christendom's clergy were teaching 100 years ago, and some have even returned to celebrating Christendom's festivals again, such as the Roman Saturnalia of December 25! Jesus and his apostles warned against such lawless ones.[10]

Translation: in other words, *when Witnesses read the Bible by itself, they find that it not only doesn't support Watchtower doctrine, but it supports historical Christianity!* Thus, this admission by the JW organization is a backhanded support to one of the principle doctrines of the Protestant Reformation: that the meaning of God's word was clear, and that it should be read by all people — without others telling them what it "really" means! The JW organization found it "strange" that individual Witnesses "reverted" to Christianity reading the Bible by itself; Martin Luther would not find it so!

To help keep other Witnesses in the fold, the organization has developed its own Bible "translation," *The New World Translation* (NWT). Biblical scholars Sakae Kubo and Walter F. Specht make short work of this translation, labeling it "the most biased" of any biblical translation.[11] The bias shows up especially in how this version portrays Jesus. Words are mistranslated, rules of grammar violated — all to keep this Bible in line with JW teachings that Jesus is not God.

> When Witnesses read the Bible by itself, they find that it supports not Watchtower doctrine, but historical Christianity!

For example, the NWT renders John 1:1 as "In the beginning the Word was, and the Word was with God, and the Word was **a** god." Here the Witnesses have added the indefinite article "a" to try to distance God from Jesus, whom they believe to be inferior to God the Father. Kubo and Specht label this twisting of Scripture as "especially objectionable," and a clear violation of Greek grammar.[12]

> As if its theological bias were not enough, the NWT's wording is wooden, at times ludicrous.

Similarly, the NWT doesn't hesitate to violate yet another rule of Greek grammar when it suits them. In Greek, as in English, when two nouns are joined by a conjunction (and), with an article before the first one, the two nouns refer to the same thing or person. For example, Titus 2:13 reads: "Looking for the blessed hope and the appearing of the glory of our great God and Savior, Christ Jesus." Paul's message is clear: Christ Jesus is both savior and God. Since such an admission would destroy JW theology, the NWT reads, "While we wait for the happy hope and glorious manifestation of the great God and of our Savior Christ Jesus." Their change is subtle, yet clearly made from theological bias — in their rendering, God and Christ Jesus are now different persons.[13]

As if its theological bias were not enough, the NWT's wording is wooden, at times ludicrous. As Old Testament scholar H.H. Rowland puts it, the NWT's translators have produced "an insult to the Word of God."[14] Like all of the Watchtower's doctrinal materials, the NWT is produced virtually anonymously, with the translators whose identities *are* known having no experience or training in Greek exegesis.[15] Thus, the reason why the Witnesses came up with a new "translation" seems apparent: they needed a Bible that was "safe" for the individual Witness to read, and that would not endanger any Jehovah's Witnesss doctrine. In any case, the translation is a sham, and any dialogues with Witnesses should employ other, *respected* translations of the Bible.

THE END OF THE WORLD

The Jehovah's Witnesses religion thrives on apocalyptic fervor. From its very beginnings with Charles Russell, up to and including today, JWs have been fascinated with the second coming of Christ, and of the terrible events that would signal His return. Sadly, the JWs went far beyond fascination and went on record as *predicting* the end of the world, and the setting up of God's "theocratic kingdom." In the 1909 edition of the Watchtower's *Studies in the Scriptures*, the organization claimed that "we consider it an established truth that the final end of the kingdoms of this world, and the full establishment of the kingdom of God, will be accomplished near the end of A.D. 1915."[16] How would the world end? In Armageddon, of course. "[T]he 'battle of the great day of God Almighty' (Rev. 16:14), which will end in A.D. 1915, with the complete overthrow of earth's present rulership, is already commenced."[17] When such prophecies failed, the JW organization, much like Big Brother in *1984*, simply went back to these books to change the embarrassing dates (the 1925 edition omits any dates from the text). Thus, the organization could claim that they never made the prophecy — but as a matter of fact, they did.

One would think that such embarrassments would be enough to make an end of all such JW "date-setting." Unfortunately, it has not. During the past century the Watchtower organization has continued to predict such apocalyptic events as 1) the resurrections of Abraham, Isaac, and Jacob in 1925 (1918, 1922, 1923); and 2) Armageddon (1941, 1968).[18] All such prophecies have failed.

Thus, despite the fact that the JW organization thinks of itself as a "prophet,"[19] they have most certainly failed the test of a true prophet as found in Deuteronomy 18: the prophesied events did not happen! One would hope that such failures would open the eyes of individual Witnesses, but such "intrusions of reality" have seldom lasted long.

> The Jehovah's Witnesses religion thrives on apocalyptic fervor.

WHAT ABOUT BLOOD TRANSFUSIONS?

> Courts have always decided in favor of involuntary transfusions of minors.

Surely one of the most well known (and at times infamous) beliefs of the Witnesses is their refusal to accept blood transfusions. It still makes newspaper headlines when a Witness dies because he or she refused a life-saving transfusion. They base this refusal on their own interpretations of Genesis 9:3-4; Leviticus 17:13-14; and Acts 15:19-21.[20] Such exegesis is utterly without foundation, and has absolutely no biblical or theological support. The passages cited above concern the *eating* of animal blood, not the transfusing of human blood.

The autocratic nature of the organization, and the control and sanctions it imposes on its members, all contribute to make this belief subject to legal and ethical scrutiny. Here, their anticulture beliefs are not only out in the open, but at times on a collision course with established laws and mores. For instance, what about the minor or incompetent JW who either refuses blood transfusion or has it refused on her behalf? Courts have rightly argued that minors or incompetents cannot make such grave decisions, nor can the parents' religious beliefs be allowed to decide the issue for their children. *Thus, courts have always decided in favor of involuntary transfusions.*[21]

In other contexts, legal outcomes have often (but not always) depended on whether or not the patient had dependents. If JW patients had dependents who would become wards of the state upon the death of their parent, courts often ordered involuntary transfusions. In the cases of adult JW patients with no dependents, courts have usually decided on behalf of the Witness, at times leading to the patient's death.

Ethically, the JWs' attitude toward blood transfusions raises a whole host of issues. A competent patient's right to refuse medical treatment is usually held inviolate. This is especially the case when such refusal is supported by the patient's religion. Still, in medical ethics, the principle of beneficence (doing good) is a valid and important one — especially when a primarily noninvasive medical procedure (blood transfusion) could save a patient's life. It is nothing less than tragic that a misuse of Scripture and poor theology can result in ruined and lost lives.

CONCLUSION

The growth of Jehovah's Witnesses worldwide mirrors the demographics of their membership. In poor, developing countries, the numbers of JWs continue to grow rapidly. In contrast, however, many countries reported fewer JWs in 1998 (e.g., France, Austria, Belgium, Denmark, Switzerland, to name five) and many others with zero growth from 1997 to 1998 (e.g., Britain, Canada, Germany, to name three).[22] In the United States, JWs grew only modestly, with the number of baptisms actually *declining*.[23]

But such anemic growth is masked by their vigorous door-to-door campaigns, and the seemingly tireless publishing. Virtually every Christian has had his or her share of "doorstop dialogues" with these eager and persistent followers of Charles Taze Russell. I have tried to show that the Jehovah's Witnesses — despite their earnestness — are touting a faith far removed from historical Christianity at several key points. I have also tried to show that, to communicate and witness effectively with JWs, making friends is more important than winning arguments. *When we live our lives as examples of Christ's love (agape), our very lives serve as*, in Peter Berger's words, *"plausibility structures" for the gospel*. That is, by our generosity, humbleness, and love, people are far more likely to listen to our message, to think it plausible. And, while such behavior may well be especially difficult when dialoguing with those such as Jehovah's Witnesses, it is not only necessary for communicating, but for living in simple obedience to the very heart of the gospel.

> When we live our lives as examples of Christ's *agape*, our very lives serve as plausibility structures for the gospel.

[1] H. Richard Niebuhr, *Christ and* Culture (New York: Harper, 1951).
[2] The following quotes are taken from Chapter 2, starting on page 45.
[3] See, for example, John 12:31; 2 Cor. 4:4.
[4] Found at http://watchtower.org/library/t22/who_rules_.htm, p. 1, 2.
[5] Heather and Gary Botting, *The Orwellian World of Jehovah's Witnesses*, (Toronto: University of Toronto Press, 1984).
[6] Ibid., pp. 49-50, 83, 86, 88, 90-91; 93, 97, 138.
[7] See Walter Martin, *Kingdom of the Cults* (Minneapolis: Bethany, 1992), p. 34.
[8] This quote is precisely the one suggested by the Watchtower when the individual Witness is accused of not believing in Jesus! See Rhodes, *Reasoning from the Scriptures*, p. 219.
[9] *The Watchtower*, September 15, 1910, p. 298.
[10] *Watchtower*, August 15, 1981, p. 29.
[11] Sakae Kubo and Walter F. Specht, *So Many Versions?* (Grand Rapids: Zondervan, 1983).
[12] Ibid., p. 99.
[13] The same bias is evident in the NWT's rendering of 2 Peter 1:1.
[14] Kubo, *So Many Versions?* p. 103.
[15] See Walter Martin and Norman Klann, *Jehovah of the Watchtower* (Minneapolis: Bethany, 1981), p. 129.
[16] "The Time Is at Hand" (Brooklyn: Watch Tower Bible and Tract Society, 1909), p. 99.
[17] Ibid., p. 101.
[18] See "One Hundred Years of Divine Direction?" (San Juan Capistrano, CA: Christian Research Institute).
[19] *The Watchtower*, April 1, 1972.
[20] See "Blood — Vital for Life," p. 6, found at http://watchtower.org/library/hb/blood_vital.htm.
[21] See, for example, *Mitchell v. Davis*, 205 SW2d 812 (Tex Civ App—Dallas, 1947).
[22] See "Statistics of Jehovah's Witnesses," p. 1, found at http://www.jwic.com/stat.htm.
[23] See the data at http://www.jwic.com/98n_amer.htm.

REFLECTING ON LESSON FIVE

1. How do you react to the JWs' rejection of many actions that we considerable normal for society? How do their attitudes compare to the concept that Christians can be in the world but not of the world?

2. In what ways can the structured society of the Watchtower religion be compared to the society Orwell depicted in the novel *1984*? How is this different from the church that Christ established?

3. Why is it important to try to make friends with those we want to win to Christ?

4. List examples from this study of terms that JWs use differently than we normally do.

5. What are the problems with *The New World Translation*? If you can acquire one, compare key passages in it with a version you know and trust.

6. What are the pitfalls of trying to set a date for Christ's return and the end of the world?

APPENDIX
THE CHURCH FATHERS AND THE DEITY OF CHRIST

The Jehovah's Witnesses attempt to portray the doctrine of the Trinity, with its corresponding stress on the deity of Christ, as a doctrine whose roots lie not in the Bible, but in paganism. They even bring in selective quotes from Christian sources (who *believe* in the Trinity, by the way) to try to prove that the Trinity doctrine is a lately manufactured belief foisted upon the people by cunning religious leaders.[1] Since the vast majority of these quoted sources *believe* both in the deity of Christ and in the Trinity, it is clear that the Jehovah's Witnesses are employing their usual habit of quoting sources out of context. But what exactly did the early church believe about Jesus? "Okay," you might say, "You've made your case regarding the New Testament, but what about those who came *after* the apostles? What about the early church? What did they believe about Christ?" The following quotes remove any doubt that the early church fathers thought Jesus Christ was God, not "a god," not Michael the Archangel, but the God of Abraham, Isaac, and Jacob.

Ignatius (A.D. 30-107) was the Bishop of the Church of Antioch, beginning in the 90s, and ending with his martyrdom in either A.D. 98 or 117. His writings contain many references to the deity of Christ. For example, "Jesus Christ our God,"[2] "by the blood of God,"[3] "the God Jesus Christ,"[4] and "our God Jesus Christ"[5] are just some of the phrases Ignatius employs regarding Jesus.

Justin Martyr (A.D. 100-165) was a famous early defender of Christianity, and wrote that Jesus, "being the first-begotten Word of God, is even God."[6] In his *Dialogue with Trypho*, he states that Jesus is ". . . deserving to be worshipped, as God and as Christ."[7]

Tatian (A.D. 110-172) was another famous defender of Christianity and wrote that "We do not act as fools, O Greeks, nor utter idle tales, when we announce that God was born in the form of a man."[8]

Irenaeus (A.D. 120-202) was a pupil of Polycarp (who himself was a pupil of the Apostle John) and he wrote that "the Father is God, and the Son is God, for whatever is begotten of God is God."[9]

Tertullian (A.D. 145-220) wrote, "Him [Jesus] we believe to have been sent by the Father into the Virgin, and to have been born of her — being both Man and God."[10]

Gregory Thaumaturgus (A.D. 205-265) wrote that "All [the persons of the Godhead] are one nature, one essence, one will, and are called the Holy Trinity; and these also are names subsistent, one nature in three persons, and one genus."[11]

Novatian (A.D. 210-280) was a Roman Presbyter who wrote that "Jesus was truly a man, but that He was also God according to the Scriptures."[12]

Athanasius (293-373) was present at the Council of Nicea in 325 and later became the Bishop of Alexandria. He wrote that Jesus "is worshipped and is believed to be God."[13]

The above list of quotations is by no means exhaustive, but is conclusive enough to prove that the deity of Christ and the doctrine of the Trinity were no "late inventions," but found in the very earliest Christian tradition. These authoritative voices demonstrate that the Church worshiped Christ as God, from the apostolic tradition all the way until the ecumenical councils formalized such belief, beginning with

Nicea in A.D. 325. J.N.D. Kelly, one of the foremost authorities on early Christian doctrine, puts it this way: "[T]he all but universal Christian conviction in the preceding centuries [before the Council of Nicea in 325] had been that Jesus Christ was divine as well as human."[14] Thus, it is the Jehovah's Witnesses' beliefs, not those of evangelical Christianity, that are opposed to the teachings of both the apostles *and* the church fathers.

---

[1] See "Should you Believe it?" found at http://watchtower.org/library/ti/should_you_believe.htm.

[2] Ignatius *Ephesians Salutation*.

[3] Ignatius *Ephesians*, 1.

[4] Ignatius *Trallians*, 7.

[5] Ignatius *Romans*, 3.

[6] Justin Martyr *Apologia*, 1, 63.

[7] Justin Martyr *Dialogue with Trypho*, 63.

[8] Tatian *Address to the Greeks*, 21.

[9] Irenaeus *Dem.*, 47.

[10] Tertullian *Against Praxeas*, 2.

[11] Gregory Thaumaturgus *On the Trinity*.

[12] Novatian *Treatise Concerning the Trinity*, 11.

[13] Athanasius *Four Discourses against the Arians*, 2.16.24.

[14] J.N.D. Kelly, *Early Christian Doctrines* (London: Adam & Charles Black, 1960), p. 138.

ADDITIONAL RESOURCES

Books

Besides the resources listed at the end of each chapter, the following books provide you with the opportunity to explore in more depth the issues of your choice regarding Jehovah's Witnesses. You may find that some of these books are now out of print. If so, check your local used bookseller, or try some of the online booksellers; they often can find hard-to-find titles.

Answering Jehovah's Witnesses: Subject by Subject by David A. Reed

Apocalypse Delayed: The Story of Jehovah's Witnesses by M. James Penton

Approaching Jehovah's Witnesses in Love: How to Witness Effectively without Arguing by Wilbur Lingle

Bible, the Christian & Jehovah's Witnesses by Gordon Lewis

Blood on the Altar: Confessions of a Jehovah's Witness Minister by David A. Reed

Crisis of Conscience by Raymond Franz

A History of Jehovah's Witnesses: From a Black Perspective by Firpo W. Carr

I Was Raised a Jehovah's Witness by Joe Hewitt

Jehovah of the Watchtower by Walter Martin and Norman Klann

The Jehovah's Witnesses, Jesus Christ, and the Gospel of John by Robert Bowman

Reasoning from the Scriptures by Ron Rhodes

Understanding Jehovah's Witnesses: Why They Read the Bible the Way They Do by Robert Bowman

We Left Jehovah's Witnesses: Personal Testimonies by Edmond Gruss

Why You Should Believe in the Trinity: An Answer to Jehovah's Witnesses by Robert Bowman

Zondervan Guide to Cults & Religious Movements ("Jehovah's Witnesses" by Robert Bowman)

Online Sources and Organizations

The author has investigated each site and has made every attempt to ensure that each is representative of historical, orthodox Christianity. But, because of the very fluid nature of web sites today, neither the author nor the publisher can guarantee that each site's doctrinal beliefs are 100 percent accurate. In addition, there are a whole host of sites that cater to ex-Jehovah's Witnesses. Some are excellent sources of information; others can be a bit mean-spirited. The reader is advised to use wisdom when searching the Internet!

Alpha and Omega Ministries. An apologetics ministry focused on the ministry of James White.
   http://www.aomin.org/

Beacon Light for Former Jehovah's Witnesses. A ministry that focuses on the needs of former Jehovah's Witnesses.
   http://www.xjw.com/

Christian Apologetics & Research Ministry. A multifaceted apologetics organization that has thorough information on cults and topics of interest to Christians.
   http://www.carm.org/index.html

Christian Research Institute. Founded by the late Walter Martin, this parachruch organization has vast resources for studying non-Christian cults and groups.
http://www.equip.org/index.html

If you are searching for a particular topic in the CRI *Journal*, then try
http://www2.omnitel.net/robertas/cri.htm

Gospel Outreach Ministries Online. A ministry that gives answers to a variety of non-Christian cults.
http://www-personal.si.umich.edu/~rlm/gomo.html

Insitute of Biblical Defense. A Pacific Northwest ministry that features apologetics, college-level courses, and outreach to the community. Deals with cults and non-Christian groups. A good site with many helpful topics and links.
http://www.biblicaldefense.org

Institute for Religious Research. An apologetics organization that analyzes modern issues and non-Christian religious groups.
http://www.irr.org/

Out of Darkness. An apologetics ministry that focuses on Jehovah's Witnesses.
http://www.outofdarkness.org/

GLOSSARY OF TERMS

Apocalypse/Apocalyptic — from the Greek "to reveal," has to do with a fascination of what must be revealed in the last days; often uses symbols and imagery.

Apologists — from the Greek word "to defend," those early Christians who defended the gospel.

Arius — a heretic from the 4th century who denied Jesus was God.

Armageddon — the scene of the famous great battle between God and the forces of evil (Rev. 16:16); Jehovah's Witnesses have (falsely) predicted this battle for over a century.

Ascension — when Jesus was "taken up" into heaven.

Blasphemy — a capital offense in Jewish Law, involves the insulting of God.

Cardinal Doctrines — those doctrines that comprise "the bottom line" of Christianity.

Eschatology — literally, the study of last things.

Grace — God's "unmerited favor" shown toward people.

Imminent Coming of Christ — Christ could come at any time; readiness is the theme.

Justification — that act by which God justifies or "makes right" the believer.

Monotheism — the belief that there is only one God.

Objective Truth — truth that remains so even if someone doesn't believe it, e.g., Salem is the capital of Oregon.

Patriarchs — the "fathers" of the Jewish faith, usually depicted as Abraham, Isaac and Jacob.

Redemption — in salvation, the act by which God has purchased ("redeemed") His people.

Sanctification — that act and/or process that occurs *after* salvation, meaning a growth in holiness.

Straw Man — in argument, claiming your opponent believes something other than he really does, then demolishing the phony or "straw" argument.

Subjective Truth — truth claims that change with the subject, e.g., "That shirt looks good on Shelley." A different observer might disagree.

Torah — Jewish for "law," usually means the first five books of the Hebrew or Christian Bibles.

Trinity — the Christian belief that the one God exists in three coeternal, codivine persons.

Yahweh — the way that the usual Hebrew personal name for God (YHWH) is spelled; because of not wanting to pronounce the sacred name, it was customary to employ the vowels from *adonai* and use them in *yhwh*, ending up with "Yahovah" or "Jahovah."

About the Author

After receiving an M.A. in Christian Apologetics from Simon Greenleaf School of Law and an M.A.R. in Religious Studies at Westminster Theological Seminary, Michael McKenzie received an M.A. and Ph.D. in Religion from the University of Southern California in 1992.

He has taught at the college level in World and Comparative Religions, Cults and Religious Groups, and Ethics. He is also the author of *Philosophy for Normal People* and *The Great Dance: A History of Religion and Politics*, as well as numerous articles and book chapters on ethical issues.

Currently, Dr. McKenzie teaches Religion, Philosophy, and Ethics at Keuka College in New York. He and his wife Allison enjoy hiking and travel — and pampering their pet cat Buzzy!